Transition Metal Doped Spintronics Materials

Dr. R. Saravanan, M.Sc., M.Phil., Ph.D.

Associate Professor
Research Centre and PG Department of Physics
The Madura College (Autonomous)
Madurai – 625 011

Published by **Materials Research Forum LLC**
Millersville, PA 17551, USA

Published as part of the book series
Materials Research Foundations
Volume 139 (2023)
ISSN 2471-8890 (Print)
ISSN 2471-8904 (Online)

Print ISBN 978-1-64490-224-0
ePDF ISBN 978-1-64490-225-7

Distributed worldwide by

Materials Research Forum LLC
105 Springdale Lane
Millersville, PA 17551
USA
http://www.mrforum.com

Printed in the United States of America
10 9 8 7 6 5 4 3 2 1

Table of Contents

Preface

Among various promising oxide candidates for spintronics applications are: wide band gap oxides based on diluted magnetic oxides, such as ZnO, TiO$_2$, SnO$_2$ doped with different transition metals Co, Mn, Fe, Cu, Ni, V, etc. and non-magnetic dopants such as N, H, etc. attract the most attention. Oxide semiconductors have advantages compared with their non-oxide counterpart. viz.,

- Wide band gap suited for applications with short wavelength light
- Transparency and dye ability with pigments
- High n-type carrier concentration
- Capability to be grown at low temperature even on plastic substrate
- Ecological safety and durability
- Low cost

The aim of the book is to synthesize and characterize various oxide based dilute magnetic spintronics materials. The distinct feature of the book is the construction of charge density of ODMS materials by deploying maximum entropy method (MEM). This charge density gives the distribution of charges in the ODMS unit cell, which is analyzed for charge related properties.

Chapter I depicts the objectives of the present book. A brief introduction to the properties of zinc oxide is given. It also highlights the oxide based semiconductors and oxide based ferromagnetic semiconductors. It gives a comparative idea about oxide based semiconductors and oxide based ferromagnetic semiconductors. It also briefs about the synthesis method employed and characterization technique involved in the present work for ODMS materials. It highlights the structural refinement technique Rietveld involved in studying the local structures. The principle behind the various characterization techniques such as, powder X-ray diffraction, scanning electron microscopy, vibrating sample magnetometry and UV Visible spectrometry which have been used for analyzing the structural properties of the crystal systems, morphological properties, magnetic properties, optical properties of the materials respectively are discussed. This chapter presents the basics of PXRD profile fitting technique and the methodology adopted for calculating the experimental charge densities. It also gives an account on the theoretical aspects involved in estimating various properties of ODMS.

Chapter II displays the results of the powder X-ray diffraction. The results of fitted PXRD profiles for all the prepared nano ferrite materials are reported. Particle size evaluation from SEM, crystallite sizes from powder X-ray profile, of the synthesized oxide based dilute magnetic semiconductors are also presented. It brings out a detailed discussion on the experimental lattice parameters, theoretical lattice parameters and oxygen positional parameters, evaluated from the results of powder XRD profile fitting and from the knowledge of atoms distribution in the zinc oxide unit cell. It discusses the x-ray diffraction analysis for various samples grown, in this present work. It discusses the variation of Bragg peaks for different concentrations. For one particular sample $Zn_{1-x}V_xO$ a vegard's plot was drawn and results were discussed.

Chapter III discusses completely the surface optical and magnetic properties of the synthesized oxide based dilute magnetic semiconductor samples. Magnetic, optical characterizations are carried out on $Zn_{1-x}Ti_xO$ (x=0.02, 0.03), $Zn_{1-x}Fe_xO$ (x = 0.02, 0.04, 0.06), $Zn_{1-x}V_xO$ (x=0.02, 0.04, 0.06), $Zn_{1-x}Ni_{x/2}V_{x/2}O$ (x=0.02, 0.04, 0.06) ODMS samples synthesized in this work.

Chapter IV displays electron density distribution studies carried out using the maximum entropy method. The results of the electron density distribution studies are presented in the form of three, two and one dimensional electron density maps.

Chapter V reveals the conclusion of the findings of the reported work.

Some findings and results published in this book have been previously published in the following research articles in highly reputed International Journals.

[1] Magnetic and optical properties of Ti Doped ZnO prepared by solid state reaction Method, T. Akilan, N. Srinivasan., R. Saravanan, Material science in semiconducting processing 30, 381 – 387, 2015. Impact factor:3.927

[2] Structural and magnetic studies on Fe doped Zinc oxide, $Zn_{1-x}Fe_xO$ synthesized by solid state reaction, T. Akilan, N. Srinivasan., R. Saravanan, Journal of Materials Science: Materials in Electronics, Springer Publication, 10854-015- 3664-1, 2015. Impact factor: 2.779

[3] Structure of Vanadium-Doped Zinc Oxide, $Zn_{1-X} V_x O$, T. Akilan, N. Srinivasan , R. Saravanan, Prasanta Chowdury, Materials and Manufacturing Processes, 29:7, 780-788, DOI: 10.1080/10426914.2014.880459, Impact factor: 4.616

[4] Structure analysis on Ni and V co-doped zinc oxide prepared by solid state reactions T. Akilan, N. Srinivasan, R. Saravanan, Journal of Materials Science: Materials in Electronics, Springer Publication, 10854-014-1957-4, 2014 . Impact factor: 2.779

Transition Metal Doped Spintronics Materials
Materials Research Foundations **139** (2023)

Materials Research Forum LLC
https://doi.org/10.21741/9781644902257

Chapter 1

Introduction

Abstract

The present work is to synthesize and characterize the diluted magnetic semiconductors. The materials studied in this work are all transition metals doped oxide semiconductors.

Keywords

Spintronics, DMS, Rietveld, Magnetic Semiconductor, Transition Metals

1.1 Objectives

The present work is to synthesize and characterize the diluted magnetic semiconductors. The materials studied in this work are all transition metals doped oxide semiconductors. The following four series have been chosen for the present work.

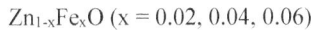

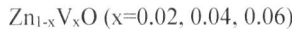

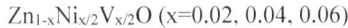

$$Zn_{1-x}Ti_xO \ (x=0.02, 0.03)$$

$$Zn_{1-x}Fe_xO \ (x = 0.02, 0.04, 0.06)$$

$$Zn_{1-x}V_xO \ (x=0.02, 0.04, 0.06)$$

$$Zn_{1-x}Ni_{x/2}V_{x/2}O \ (x=0.02, 0.04, 0.06)$$

The main objectives of the present work,

- To prepare ZnO semiconductors by solid-state reaction method (SSR).

- To investigate the prepared transition metal doped dilute oxide semiconductor by powder X-ray diffraction for the detailed structural analysis.

- To analyze the morphology of the semiconductors by scanning electron microscopy (SEM).

- To analyze the elemental compositions of the prepared materials using energy dispersive X-ray spectroscopy (EDS).

- To estimate the optical band gap (E_g) for all the prepared materials by UV-visible absorption spectra by using Tauc plot technique (Wood and Tauc, 1972).

- To analyze the magnetic properties of the materials by vibrating sample magnetometer (VSM).

- To study charge density distribution for the above mentioned transition metal doped dilute oxide semiconductor using Rietveld (Rietveld, 1969) and maximum entropy method (MEM) (Collins, 1982) using powder X-ray diffraction data.

- To correlate the charge derived properties of the transition metal doped dilute magnetic oxide semiconductors with the addition of the dopant with the experimental results.

1.2 Crystal structure of ZnO

At ambient pressure and temperature, ZnO crystallizes in the wurtzite structure, as shown in figures 1.1. This is a hexagonal lattice, belonging to the space group *P63mc* with lattice parameters a=b=3.2495 Å and c=5.2062 Å and characterized by two interconnecting sublattices of Zn^{2+} and O^{2-}, such that each Zn ion is surrounded by tetrahedral O ions and vice versa. This tetrahedral coordination gives rise to polar symmetry along the hexagonal axis. This polarity is responsible for a number of the properties of ZnO including its piezoelectricity and spontaneous polarization, and is also a key factor in crystal growth, etching and defect generation. A detailed schematic arrangement of atoms in the conventional unit cell is shown in figure 1.1. This wurtzite lattice is composed of two interpenetrating hexagonal close-packed (hcp) sub-lattices, each of which consists of one type of atom displaced with respect to each other along the threefold c-axis. From this figure it is clearly seen that every atom of one kind (e.g. Zn) is surrounded by four atoms of the other kind (O) or vice versa, which are positioned at the edges of a tetrahedron. This tetrahedral coordination is typical of sp^3 covalent bonding. The Zn-O bond length is 1.992 Å in the direction parallel to the c-axis of the hexagonal unit cell and 1.973 Å in the other three directions of the tetrahedral arrangement. In a wurtzite lattice, there are lattice parameters a, b and c and the internal parameter u and bond angles α and β=109.47°. The internal parameter u is defined as the length of the bond parallel to the c-axis (anion-cation bond length or the nearest-neighbour distance) divided by the lattice parameter c. The lattice constants of the ZnO unit cell are a=b=3.2495 Å and c=5.2062 Å, yielding a c/a ratio of 1.602, which is close to the ideal value of 1.633 expected for the hcp unit cell. In addition to intrinsic material properties, the lattice parameters are affected by extrinsic properties such as the free electron concentration (via the deformation potential of the conduction band minimum), the concentration of foreign impurities with different ionic radii which can replace the host atom, substrate-induced strain and temperature (Kucheyev et al., 2002; Catti et al., 2003).

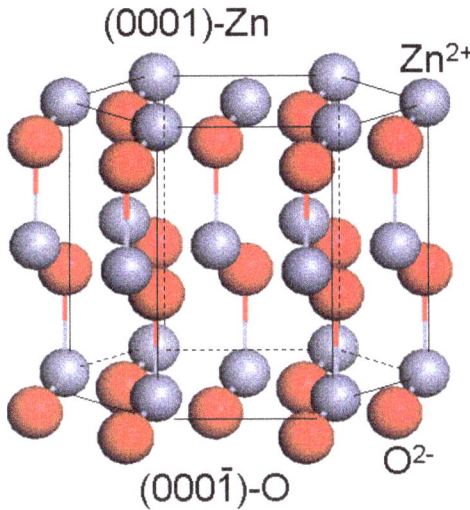

Figure 1.1 ZnO wurtzite crystal structure

1.3   Mechanical properties

ZnO is a relatively very soft material with approximate hardness just 4.5. Its elastic constants are relatively smaller than those of other III-V semiconductors, e.g. GaN. The high heat capacity and high heat conductivity, low values of thermal expansion and high melting points are some of the characteristics of ZnO. ZnO has been proposed to be a more promising UV emitting phosphor than GaN because of its larger exciton binding energy (60 meV). Among the semiconductors bonded tetrahedrally, it is found that ZnO has the highest piezoelectric tensor. This makes it an important material for many piezoelectric applications, which require a high degree of electromechanical coupling among them (Baomei et al., 2008).

1.4   Electrical properties

The fundamental study of the electrical properties of ZnO nanostructures is crucial for developing any future applications in nanoelectronics. ZnO has a quite large band gap of 3.3 eV at room temperature. The advantages of a large band gap include higher values of breakdown voltages, sustaining large electric fields, high-temperature and high-power operations. ZnO has n-type character, in the absence of doping. Non-stoichiometry is usually the origin of n-type character. Due to defects such as oxygen vacancies and zinc interstitials, ZnO nanowires are reportedly show n-type semiconductor behavior. The main impediment of ZnO for wide-ranging applications in electronics and photonics rests with the diffculty of p-type doping. Successful p-type doping for ZnO nanostructures will greatly enhance their future applications in nanoscale electronics and optoelectronics. P-type and n-type ZnO nanowires can serve as p-n junction diodes and light emitting diodes (LED) (Baomei et al., 2008).

1.5 Optical properties

Zinc oxide is generally transparent to visible light but strongly absorbs ultraviolet light below 3655 Å. The absorption is typically stronger than any other white pigments. In the region of visible wavelengths, regular zinc oxide appears white, but rutile and anatase titanium dioxide have a higher refractive index and thus has a superior opacity. The band gap energy (between valence and conducting bands) is 3.2 eV, this corresponds to the energy of 3655 Å photons. Under ultraviolet light zinc oxide is photoconductive. The combination of optical and semiconductor properties make doped zinc oxide a contender for new generations of devices. Solar cells require a transparent conductive coating, indium tin oxide and doped zinc oxide are the best materials. Intrinsic optical properties of ZnO nanostructures are being intensively studied for implementing photonic devices. Photoluminescence (PL) spectra of ZnO nanostructures have been extensively reported. Excitonic emissions have been observed from the photoluminescence spectra of ZnO nanorods. It is shown that quantum size confinement can significantly enhance the exciton binding energy. Strong emission peak at 380 nm due to band-to-band transition and green-yellow emission band related to oxygen vacancy are observed. PL spectra show that ZnO nanowire is a promising material for UV emission, while its UV lasing property is of more significance and interest. Due to its near-cylindrical geometry and large refractive index (~2.0), ZnO nanowire/nanorod is a natural candidate for optical waveguide. The additional advantages of ZnO nanowire lasers are that the excitonic recombination lowers the threshold of lasing, and quantum confinement yields a substantial density of states at the band edges and enhances radiative efficiency. Optical waveguiding using dielectric nanowire also achieved considerable progress. Recently, ZnO nanowires were reported as sub-wavelength optical waveguide. Optically pumped light emission was guided by ZnO nanowire and coupled into SnO_2 nanoribbon. These findings show that ZnO nanostructures can be potential building blocks for integrated optoelectronic circuits (Baomei et al., 2008).

1.6 Photoluminescence properties of ZnO

At room temperature, the UV emission band is related to a near band-edge transition of ZnO, namely, the recombination of the free excitons. The broad emission band literally between 420 nm and 700 nm observed nearly in all samples regardless of growth conditions is called deep level emission band (DLE). The DLE band has previously been attributed to several defects in the crystal structure such as O-vacancy (V_O), Zn-vacancy (V_{Zn}), O-interstitial (O_i), Zn-interstitial (Z_{ni}), and extrinsic impurities such as substitutional Cu. Recently, this deep level emission band had been identified and at least two different defect origins (V_O and V_{Zn}) with different optical characteristics were claimed to contribute to this deep level emission band (Zhao et al., 2005; Borseth et al., 2006; Klason et al., 2008).

At low cryogenic temperatures, bound exciton emission is the dominant radiative channel. The luminescence spectrum from ZnO extends from the band edge to the green/orange spectral range. Very common is a broad band centered about 2.45 eV extending from the blue into the green range.

1.7 Diluted magnetic semiconductor

Semiconductor spintronics is one of the most perspective directions for the modern technology development. The crucial point to ensure creation of semiconductor spintronics devices is the development of appropriate materials, having a ferromagnetic ordering at room temperature and high values of magnetization. Diluted magnetic semiconductors (DMSs) are the potential candidates to be used in semiconductor spintronics. Since Dietl et al. (Dietl et al., 2000) proposed the idea of ferromagnetism introduced in the semiconductors by the presence of magnetic cations like Mn doped in semiconductor matrix, there has been a number of reports of these doped semiconductors (Dietl et al., 2000; Mallick et al 2009). Diluted magnetic semiconductors (DMSs) formed by substituting the cations of III-V or II-VI nonmagnetic semiconductors by ferromagnetic Mn, Fe, Co and Ni exhibit a number of unique, magneto-optical and magneto-transport properties, pertinent for magneto-electronic and spintronic devices (Pan et al., 2007; Rath et al., 2009). Room temperature ferromagnetism in transition metal doped ZnO has been continually envisaged by numerous groups using variety of theoretical and experimental studies.

There is a wide class of semiconducting materials which is characterized by the random substitution of a fraction of the original atoms by magnetic atoms. The materials are commonly known as semimagnetic semiconductors (SMSC) or diluted magnetic semiconductors (DMS). The most common SMSC are II-VI compounds (like CdTe, ZnSe, CdSe, CdS, etc.), with transition metal ions (e.g. Mn, Fe or Co) substituting their original cations. There are also materials based on IV-VI (e.g. PbTe, SnTe) and recently III-V (e.g. GaAs, InSb) crystals. On the other hand, rare earth elements (e.g. Eu, Gd, Er) are also used as magnetic atoms in SMSC. These mixed crystals (semiconductor alloys) may be considered as containing two interacting subsystems. The first of these is the system of delocalized conduction and valence band electrons. The second is the random, diluted system of localized magnetic moments associated with the magnetic atoms. The fact that both the structure and the electronic properties of the host crystals are well known, they are perfect for understanding the basic mechanisms of the magnetic interactions and the localized spins of magnetic ions. The coupling between the localized moments results in the existence of different magnetic phases (such as paramagnets, spin glasses and antiferromagnets). The wide variation of both host crystals and magnetic atoms provides materials which range from wide gap to narrow band gap semiconductors. Several of the properties of these materials may be tuned by changing the concentration of the magnetic ions.

DMS material exhibit giant Faraday rotation. The change in polarization of light on interaction with magnetic field results in extremely large Zeeman splitting. The spin state of the Mn ion is mediated to neighbor sites by propagating spin-polarized carriers which tend to align the whole ensemble of the localized spins. The dependence of the energy of the system on the relative orientation of Mn moments is generally referred to as an exchange interaction (Jungwirth, 2006). Hideo ohno et al., at the Tohuku University were the first to measure ferromagnetism in transition metal doped compound semiconductors such as indium arsenide (InMnAs) and gallium arsenide doped with manganese (GaMnAs). These materials exhibit reasonably high Curie temperatures well below the room temperature with the variation in concentration of p-type charge carriers for various semiconductor hosts doped with different transition atoms (Ohno et al., 1998).

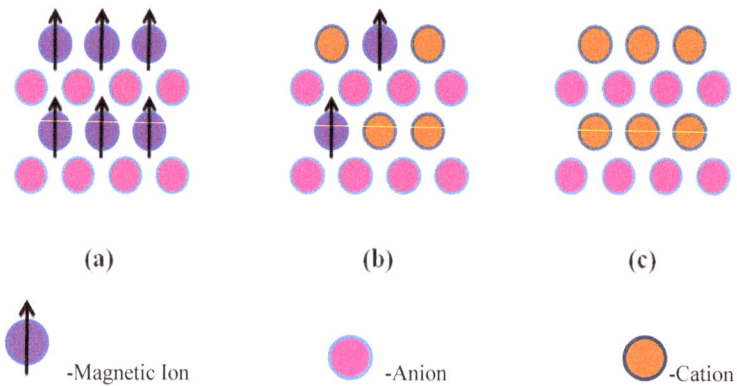

(a)   (b)   (c)

↑ -Magnetic Ion   ● -Anion   ● -Cation

Figure1.2 Three types of semiconductors (a) magnetic semiconductor, (b) diluted magnetic semiconductor, (c) nonmagnetic semiconductor

1.7.1 Oxide based diluted magnetic semiconductor

Most conducting materials are opaque to light at visible wavelengths. On contrary, the wide band gap oxide semiconductors such as ZnO, TiO_2, In_2O_3, etc. are being highly transparent in visible region and also good conductors due to the presence of native defects (King et al., 2011). They can be doped heavily by n-type carriers. Among various promising oxide candidates for spintronics applications are: wide band gap oxides based on diluted magnetic oxides, such as ZnO, TiO_2, SnO_2 doped with different transition metals Co, Mn, Fe, Cu, Ni, V, etc. and non-magnetic dopants such as N, H, etc. attract the most attention. Oxide semiconductors have advantages compared with their non-oxide counterpart. viz.,

- Wide band gap suited for applications with short wavelength light
- Transparency and dye ability with pigments
- High n-type carrier concentration
- Capability to be grown at low temperature even on plastic substrate
- Ecological safety and durability
- Low cost, etc. In addition, large electronegativity of oxygen is expected to produce strong p-d exchange coupling between band carriers and localized spins (Mizokawa et al., 2002).

Furthermore, the magnetic properties of oxides span from diamagnetic, paramagnetic to ferromagnetic and antiferromagnetic, and their chemical reactivity can be tailored from being highly reactive to inert. All these promising properties originate from the strong and competitive exchange coupling between charge, orbital, lattice and spin degrees of freedom in these oxide materials. Hence ZnO and TiO_2 are employed as host semiconductors.

The major problem in DMOS's to practical application is their weak magnetic signal as the doping concentration is usually a fractional percent. Yan et al., proposed the alternative solutions such as preparation of concentrated oxide ferromagnetic semiconductors (COFSs) by sputtering of thin (less than 1 nm) ferromagnetic metal (Co, Fe) layers and oxide (ZnO, TiO_2, In_2O_3, etc.) layers. These COFSs are one of the promising candidates for spintronics application (Zhang et al., 2006).

1.8 Methodology

1.8.1 Structural determination using Rietveld refinement technique

To understand the properties of materials, it is essential to know about their atomic structure. The X-ray diffraction methods are the best techniques to study the atomic structure of the materials. In the past three decades, powder diffraction has played a vital role in the field of structural physics, chemistry and material science. High temperature superconductors and high pressure research have relied mostly on powder diffraction techniques. In early 1990s, powder diffraction data have been used for the structural determination of only very few crystals. But today, numerous organic and inorganic crystal structures have been solved through powder diffraction data. Many powder diffraction methodologies have developed which have contributed to analyze the structure of the compounds. Since the powder diffraction peaks grossly overlap in the conventional powder diffraction method, it hinders the exact determination of the crystal structure. The Rietveld method (Rietveld, 1969) minimizes the impact of these overlapping peaks and determines the real crystal structure. Rietveld method (Rietveld, 1969) is a technique for the crystal structure refinement which uses the whole powder diffraction pattern. In Rietveld method (Rietveld, 1969), least-squares approach is used so that the calculated and measured diffraction profiles are optimized and by iterative technique, the profiles are refined. The following sections explain Rietveld (Rietveld, 1969) refinement and several powder profile parameters deployed in the technique.

1.8.2 Rietveld method

The Rietveld (Rietveld, 1969) analysis is a unique and valuable method for extracting detailed structural information from X-ray powder data. Rietveld method (Rietveld, 1969) was devised to quantify the structural treatment, doping, compression and strain. This refinement can be performed on the obtained experimental powder X-ray diffraction data. The method was introduced by Hugo Rietveld in 1969 (Rietveld, 1969). It is a whole profile structure refinement method based on least-square fitting. A calculated profile fit of the structure is matched with experimental data and then the selected parameters are refined repeatedly in order to get the best fit between the calculated and experimental profiles.

The Rietveld method (Rietveld, 1969) is a technique for refining structural parameters and lattice parameters directly from whole powder diffraction data sets. The structural parameters are fractional coordinates, occupation factors, isotropic/anisotropic atomic displacement parameters, etc. This method was initially only applied to neutron data due to their simple peak shape. The Rietveld refinement (Rietveld, 1969) uses different peak-shape functions such as Lorentzian and pseudo-Voigt. It should be mentioned that successful structural analysis by this method is directly related to the quality of powder diffraction data. The structural refinement gives model of the crystal structure. This model has approximations on unit cell dimensions and atomic coordinates and it should have the same space group. The refinement will be satisfactory when the best fit between the observed and the calculated model is achieved. The Rietveld method (Rietveld, 1969) uses a least-squares approach, to minimise the weighted sum of the point-by-point differences squared in a powder diffraction pattern. The intensity is calculated using the experimental parameters such as instrumental parameters, background intensities, absorption, extinction etc. and the sample dependent parameters such as unit cell parameters, atomic fractional coordinates atomic occupancy of each crystallographic site, Debye-Waller factors and background. Minimization (M) is defined by an equation such as,

$$M = \sum_i w_i \left(y_{i\,(obs)} - y_{i\,(calc)}\right)^2 \qquad \qquad .... (1.1)$$

w_i is weighting factor for each observed point and is equal to $w_i = 1/y_i$. $y_{i\,(obs)}$ and $y_{i\,(calc)}$ are observed and model calculated intensities at each step for i data points. JANA 2006 is the versatile software deployed for structural analysis and refinement of the samples (Petříček et al., 2006). Constraints can be applied to reduce the degree of freedom of the set of equations that has to be solved. The quality of the Rietveld refinement (Rietveld, 1969) is indicated by some residual functions. The profile R-factor, which is the most straightforward discrepancy index, is a measure of the disagreement between the observed and calculated profile. The quality of the fit can be monitored by the R_p (profile factor), and R_{wp} (weighted profile factor), the goodness of fit is indicated by the χ^2 value, which is optimal when $\chi^2=1$, representing an ideal match between experimental and theoretical data, which is given as:

$$R_p = \frac{\sum y_{i(obs)} - y_{i(cal)}}{\sum y_{i(obs)}} \qquad \qquad .... (1.2)$$

$$R_{wp} = \frac{\sum w_i (y_{i(obs)} - y_{i(cal)})^2}{\sum w_i (y_{i(obs)})^2} \qquad \qquad .... (1.3)$$

$$\chi^2 = \frac{\sum w_i (y_{i(obs)} - y_{i(cal)})^2}{N - P} \qquad \qquad .... (1.4)$$

N - number of data points

P - number of refined parameters

1.8.3   Electron density distribution

The quantum mechanical theory explains that the electron density is the measure of the probability of an electron being present at a specific location. The atoms are surrounded by electron clouds. The electron density is defined as the number of electrons per unit volume. The chemical bondings as well as the physical and chemical properties of the crystal systems have been analyzed by the electron density of a system. The electron density distribution study is applied in many disciplines in chemistry, physics, biology and geology

(Stout et al., 1989). The study of chemical bonding and internal local structure of a crystalline system is very important and it gives useful information about the transport properties which can be effectively utilized for device applications. For the precise understanding of the nature of chemical bonds, it is essential to study about the electron density distribution between the atoms.

Since the lattice has a periodicity, the electron density is also considered to behave as a periodic function. The number of electrons in any volume element dV is $\rho(x, y, z)dV$. In a X-ray scattering experiment, the wavelet scattered by this element is

$$\rho(x, y, z)\exp(-2\pi i(hx + ky + lz))dV \qquad \text{.......... (1.5)}$$

The resultant sum of contributions from all the elements in the unit cell, i.e., the integral over its volume gives

$$F_{hkl} = \int \rho(x, y, z) \exp(-2\pi i(hx + ky + lz))dV \qquad \text{.......... (1.6)}$$

The structure factor is considered as a resultant of adding the scattered waves in the direction of the *hkl* reflection from the atoms in the unit cell. This approach was based on the assumption that the scattering power of the electron cloud surrounding each atom could be equated to that of the proper number of electrons concentrated at the atomic centre. But the structure factor may equally well be considered as the sum of the wavelets scattered from all the infinitesimal elements of electron density in a unit cell, with no assumptions being made about the distribution of this density. The electron density $\rho(r)$ is defined as the number of electrons per unit volume.

The geometric properties of unit cells can be deduced from the locations of reflections on various kinds of X-ray diffraction photographs. It concerned with the measurement of the relative intensities of these reflection, since it is from the intensities that it hopes to be able to deduce the electron density distribution in the crystal cell. There are connections between the intensities and the electron density distribution. There are a few general precautions, which are applicable to any intensity measuring method. If the structure factors and phases are known, the electron-density distribution of the unit cell can be calculated. The magnitude of individual structure factors are calculated as the square-root of the measured diffraction intensity and their phases are determined by solving the structure. The interpretation is described as a model, which is improved by least-squares refinement based on the structure factors. The electron density can then be calculated as a Fourier summation of phased structure factors.

Intensities of diffracted X-rays are due to interference effects of X-rays scattered by all the different atoms in the structure. The diffraction pattern is the Fourier transform of the crystal structure, corresponding to the pattern of waves scattered from an incident X-ray beam by a single crystal; it can be measured by experiment (only partially, because the amplitudes are obtainable from the directly measured intensities via a number of correction, but the relative phases of the scattered waves are lost), and it can be calculated (giving both amplitudes and phases) for a known structure. So, the crystal structure is the Fourier transform of the diffraction pattern and is expressed in terms of electron density distribution concentrated in atoms; it cannot be measured by direct experiment, because the scattered X-rays cannot be refracted by lenses to form an image as done with light in an optical microscope, and it cannot be obtained directly by calculation, because the required relative phases of the waves are unknown. We can calculate the electron density distribution given a set of structure factors, using the Fourier series.

1.8.3.1 Fourier method

Any well-behaved function can be represented by means of suitable series of trigonometric terms called Fourier series. We can imagine the unit cell is divided into small volumes dV in which there are $\rho(r)$ dV number of electrons. The scattered amplitude from such a small volume will be $\rho(r)$ dV times as much that from an electron at the same position. From this we find the total scattered amplitude from the distribution of electron density $\rho(r)$. F(H) can be expressed in terms of density $\rho(r)$ as

$$F(H) = \int \rho(r) \exp(2\pi i H \cdot r) dV \qquad \text{........ (1.7)}$$

The inverse Fourier transform of this gives the electron density

$$\rho(r) = \int \rho(x,y,z) \exp(-2\pi i H \cdot r) dV = \frac{1}{V} \sum F(H) \exp(-2\pi i H \cdot r) \qquad \text{....... (1.8)}$$

Since F(H) is defined at the discrete set of reciprocal lattice points k, the integral is replaced by the summation. Writing the structure factor as F(H) = A(H) + iB(H), then

$$\rho(r) = \frac{1}{V} \sum ((A(H) + iB(H)) (\cos(2\pi H \cdot r) - i\sin(2\pi H \cdot r)) \qquad \text{........ (1.9)}$$

where A(H) = A(-H) and B(H) =B(-H) since the electron density is a real function.

Therefore, $\rho(r) = \frac{1}{V} \sum_{1/2} (2A(H)\cos(2\pi H \cdot r) + 2B\sin(2\pi H \cdot r)) \qquad \text{......... (1.10)}$

with A(H) = $|F(H)|\cos\varphi$ and B(H) = $|F(H)|\sin\varphi$ resulting in

$$\rho(r) = \frac{2}{V} \sum_{1/2} (|F(H)|\cos\varphi\cos(2\pi H \cdot r) + |F(H)|\sin\varphi \sin(2\pi H \cdot r)) \qquad \text{......... (1.11)}$$

which reduces to

$$\rho(r) = \frac{2}{V}\Sigma_{1/2}(|F(H)|cos(2\pi H \cdot r) - \varphi(H))$$
.... (1.12)

In other words, each structure factor contributes a plane wave to the total density with wave vector H and phase φ. As we know the formation of the image, which is the density, requires knowledge of the phases of the structure factors. Once an approximation to the scattering density is known, $\varphi(H)$ may be calculated on the basis of this approximation, and an admittedly imperfect image of the structure can be obtained. At the same time, anomalous scattering can be corrected for which can be done by subtracting the calculated contributions $\Delta A_{calc}^{anamalous}$ and $\Delta B_{calc}^{anamalous}$ from A and B, respectively, using the anomalous scattering factors and f' and f''.

The period of the plane wave with amplitude F(H), in the direction of the wave vector H, equals 1/H. The period is therefore shorter for higher-order reflections are included in the summation, the resolution of the image improves. The improvement is analogous to the increase in resolution in an optical image obtained with shorter-wavelength radiation.

The non-existence of lenses for X-ray beams makes it necessary to use computational methods to achieve the Fourier transform of the diffraction pattern into the image. In the calculation of charge density by this method we need infinite number of Fourier co-efficient to perform the Fourier Synthesis. But we use only a limited number of Fourier co-efficient and ignore experimental errors by setting all the missing Fourier co-efficient as zero. We neglect the missing structure factors by setting them to zero simply because the experiment cannot be or was not carried out. This is a highly biased assumption. This results in the unphysical negative electron density and hampers the use of it in understanding the finite details like the bonding charge in valence region.

1.8.3.2 Maximum entropy method

Experimental electron density distribution can be reconstructed from accurate X-ray diffraction data and analysis steps (Iversen et al., 1996). Maximum entropy method (MEM) (Collins, 1982) is used to reconstruct the charge density distributions in the molecular systems. To understand the physical and chemical properties of the material systems, one should require the knowledge about their charge distribution (Bader et al., 1991). The resultant density distribution gives detailed information about the structure, without using the structural model. MEM (Collins, 1982) electron density map is an accurate mathematical tool for structural analysis. Compared to the map drawn by conventional Fourier transformation, the resolution of MEM electron density map is higher (Sakata et al., 1990). MEM uses the structure factors retrieved from Rietveld (Rietveld, 1969) refinement. Hence, the combination of MEM (Collins, 1982) and Rietveld (Rietveld, 1969) method provide the detailed structure model. Using XRD, it is impossible to collect the exact values of all the structure factors. Hence, there are some errors in the number of observed structure factors by the experiment. The uncertainties due to these errors must be rectified. MEM (Collins, 1982) introduces the concept of entropy to tackle the uncertainty properly. So, the concept behind maximum entropy method (MEM) (Collins, 1982) is to obtain the electron density distribution which is consistent with the observed structure factors and to leave the uncertainties minimum. The mathematical description of MEM (Collins, 1982) is explained below.

The maximum entropy method (MEM) is an information-theory-based technique to enhance the information and was developed in the field of radio astronomy to enhance the information obtained from noisy data (Gull and Daniell, 1978). The theory is based on the same equations that are the foundation of statistical thermodynamics. Both the statistical entropy and the information entropy deal with the most probable distribution. In the case of statistical thermodynamics, this is the distribution of the particles over position and momentum space, while in the case of information theory, the distribution of numerical quantities over the ensemble of pixels is considered. The probability of a distribution of N identical particles over m boxes, each populated by n_i particles, is given by

$$P = \frac{N!}{n_1! n_2! n_3! \dots n_m!} \qquad \dots\dots\dots (1.13)$$

As in statistical thermodynamics, the entropy is defined as ln P. Since the numerator is constant, the entropy is, apart from a constant, equal to

$$S = - \sum_i n_i \ln n_i \qquad \dots\dots\dots (1.14)$$

where Stirlings' formula ((ln N! =N ln N - N) has been used.

In case there is a prior probability q_i for box i to contain n_i particles, then, becomes

$$P = \frac{N!}{n_1! n_2! n_3! \dots n_m!} \times q_1^{n_1} q_2^{n_2} \dots \dots q_m^{n_m} \qquad \dots\dots\dots (1.15)$$

which gives, for the entropy expression,

$$S = - \sum_i n_i \ln n_i + \sum_i n_i \ln q_i = - \sum_{i=1}^{m} n_i \ln \frac{n_i}{q_i} \qquad \dots\dots\dots (1.16)$$

The maximum entropy method was first introduced into crystallography by Collins (Collins, 1982), who based on equation (1.16), expressed the information entropy of the electron density distribution as a sum over M grid points in the unit cell, using the entropy formula (Jaynes, 1968).

where both $\rho'(r)$ and $\tau'(r)$ are actual electron density and prior probability of actual electron density in a unit cell respectively as

$$\rho'(r) = \frac{\rho(r)}{\sum_r \rho(r)} \quad \text{and} \quad \tau'(r) = \frac{\tau(r)}{\sum_r \tau(r)} \qquad \text{......... (1.18)}$$

where $\rho(r)$ and $\tau(r)$ are the electron density and prior electron density at a fixed r in a unit cell respectively. In the present theory, the actual densities are treated instead of normalized densities and $\rho'(r)$ become $\tau'(r)$ when there is no information. The $\rho'(r)$ and $\tau'(r)$ are normalized as

$$\sum \rho'(r) = 1 \quad \text{and} \quad \sum \tau'(r) = 1 \qquad \text{......... (1.19)}$$

The entropy is maximized subject to the constraint

$$C = \frac{1}{N} \sum_k \frac{|F_{cal}(k) - F_{obs}(k)|^2}{\sigma^2(K)} \qquad \text{......... (1.20)}$$

Where N is the number of reflections used for MEM analysis, $\sigma(k)$, standard deviation of $F_{obs}(k)$, the observed structure factor and $F_{cal}(k)$ is the calculated structure factor given by

$$F_{cal}(k) = V \sum_r \rho(r) exp(-2\pi i k.r) \, dV \qquad \text{......... (1.21)}$$

where V is the volume of the unit cell. The constraint C is known as weak constraint, in which the calculated structure factors agree with the observed structure factors as a whole when C becomes unity. Equation (1.21) shows, the structure factors are given by the Fourier transform of the electron density distribution in a unit cell. In the MEM (Collins, 1982) analysis, there is no need to introduce the atomic factors, by which the structure factors are normally written. It should be emphasized here that it would be an assumption to use the atomic form factors in the formulation of the structure factors. Equation (1.21) guarantees that it is possible to introduce any kind of deformation of the electron densities in real space as long as information concerning such a deformation is included in the observed data.

We use Lagrange's method of undetermined multiplier (λ) in order to constrain the function C to be unity while maximizing the entropy. We then have

$$Q = S - \frac{\lambda}{2} C \qquad \text{......... (1.22)}$$

Materials Research Forum LLC
https://doi.org/10.21741/9781644902257

$$Q = -\sum \rho'(r) ln\left(\frac{\rho'(r)}{\tau'(r)}\right) - \frac{\lambda}{2N}\sum_k \frac{|F_{cal}(k)-F_{obs}(k)|^2}{\sigma^2(k)} \qquad \dots\dots (1.23)$$

And when $\dfrac{dQ}{d\rho} = 0$ and using the approximation,

$ln\, x = x - 1$ we get,

$$\rho(\boldsymbol{r_i}) = \tau(\boldsymbol{r_i})exp\left\{\left(\frac{\lambda F_{000}}{N}\right)\left[\sum\frac{1}{\sigma(H)^2}\right]|F_{obs}(\boldsymbol{k}) - F_{cal}(\boldsymbol{k})|exp(-2\pi j\,\boldsymbol{k.r})\right\} \qquad \dots (1.24)$$

where $F_{000} = Z$ is the total number of electrons in a unit cell.

In order to solve equation (1.24) an approximation is introduced to replaces $F_{cal}(\boldsymbol{k})$ as

$$F_{cal}(\boldsymbol{k}) = V\sum \tau(r)exp(-2\pi i\,\boldsymbol{k.r})\,dV \qquad \dots\dots (1.25)$$

This approximation can be called zeroth order single pixel approximation. By using this approximation the right hand side of equation (1.24) becomes independent of $\tau(\boldsymbol{r})$ and equation (1.24) can be solved in an iterative way starting from a given initial density for the prior distribution. A uniform density distribution is given as the prior density.

$$\tau'(\boldsymbol{r}) \le \tau(\boldsymbol{r}) \ge \frac{Z}{M} \qquad \dots\dots (1.26)$$

where M is the total number of pixels for which the electron density is calculated. In the calculation of $\rho(\boldsymbol{r})$, all of the symmetry recruitments are satisfied, and the number of electrons (Z) is always kept constant through an iteration process. Mathematically, the summation concerning $\rho(\boldsymbol{r})$ in the above equations should be written as an integral. Since we must use a very limited number of pixels in the numerical calculation, the integral is replaced by the summation in the above equations (equations 1.14 to 1.26).

After completion of the MEM (Collins, 1982) enhancement, it becomes possible to evaluate the reflections missing from the summation. In a Fourier summation, the amplitudes of the unobserved reflections are assumed to be equal to zero, while the MEM (Collins, 1982) technique provides the most probable values. When extinction is present in the data set, it must be corrected before the MEM procedure is started. The structure factors must similarly be corrected for anomalous scattering, if present. Both corrections require a model for their evaluation. The independent-atom model is usually adequate for this purpose. The advantage of maximum entropy method is a statistical deduction that can yield a high resolution density distribution from a limited number of diffraction data without using a structural model. It has been suggested that MEM (Collins, 1982) would be a suitable method for examining electron densities in the inner atomic region, for example, bonding region. It gives less biased information on the electron densities as compared to conventional Fourier synthesis.

1.8.3.3 Methodology for the determination of charge density

The technological advances in recent years bring demands for integrated three dimensional visualization systems to deal with both structural models and volumetric data, such as electron and nuclear densities. The crystal structures and spatial distribution of various physical quantities obtained experimentally and by computer simulations should be understood three-dimensionally. Once the structure factors are refined, they are further utilized for the evaluation of MEM (Collins, 1982) charge density. The maximum entropy method (Collins, 1982) gives information on reconstructing the structure factor using preliminary information like position, type, space group, etc. The calculated structure factor is then compared with the observed one and the resultant calculated structure factor and observed one, which is used for the reconstruction of charge density using MEM analysis (Collins, 1982). The electron density distributions in the unit cell are constructed through the PRIMA (Practice of Iterative MEM Analyses) (Izumi and Dilanian, 2002) software. The input file contains the cell parameters, space group, pixels, total charge, Lagrange parameter and structure factors.

In the present work, the unit cell was divided into $64 \times 64 \times 64$ pixels and the initial electron density at each pixel was fixed uniformly as Z/a_0^3, where Z is the number of electrons in the unit cell. The electron density is evaluated by carefully selecting the Lagrange multiplier in each case such that the convergence criterion C becomes unity after performing minimum number of iterations. The three dimensional (3D) electron density was plotted using VESTA (Visualization of Electronic and Structural Analysis) (Momma and Izumi, 2006) software package. VESTA (Momma and Izumi, 2006) software deals with structural models and volumetric data at the same window. To understand the nature of bond in the materials, two dimensional (2D) and one dimensional (1D) distribution of electron densities on different lattice plane have been mapped and discussed in the following chapters.

1.9   Characterization techniques

1.9.1   Powder X-ray diffraction

To understand the structure and properties of a material, one should know how the atoms are arranged in the crystal structures. The spatial arrangement of atoms in the material can be studied through diffraction experiments. In an experiment of diffraction, the incident waves strike on the material and a detector records the outgoing diffracted waves. The scattered waves with constructive interference produce a diffraction pattern. XRD is the science of determining the arrangement of the atoms within a crystal from the manner in which a beam of X-rays is scattered from the electrons within the crystal. Till 1895 the study of matter at the atomic level was a difficult task but the discovery of electromagnetic radiation with 1.0 Å wavelength, appearing at the region between gamma-rays and ultraviolet, makes it possible. As the atomic distance in matter is comparable with the wavelength of X-ray, the phenomenon of diffraction finds its way through and gives many profitable results related to the crystalline structure. X-ray diffraction (XRD) is a versatile, non-destructive technique used for qualitative and quantitative analysis of a crystalline materials. Crystals with perfect periodicity have sharp diffraction peaks whereas crystals with less periodicity have broadened or distorted diffraction peaks. The structure of amorphous materials can also be studied through diffraction peaks. This experimental technique has been used to determine the overall structure of bulk solids, including lattice constants, identification of unknown materials, orientation of single crystals, orientation of polycrystalline, stress, strain, texture, films thickness *etc*.

XRD analysis can provide information related to the crystalline quality and electron densities of crystallographic planes of the given material. In addition, the diffraction spectrum also provides information regarding the types of phases present in the material and can be used to calculate the approximate average grain size using Scherrer's equation.

If a beam of X-ray interacts with two successive crystallographic planes of given Miller indices, the path difference (2d) between two reflected beams is equal to $2d_{hkl} \sin \theta$ shown in figure 1.3. The interference is constructive when the phase shift is a multiple of 2π; this condition can be expressed by Bragg's law (Bragg, 1913):

$$n\lambda = 2d \sin \theta \qquad\qquad\qquad \text{.......... (1.27)}$$

where n is an integer determined by the order given, λ is the wavelength of the X-ray, d is the spacing between the planes in the atomic lattice and θ is the angle between the incident ray and the scattering planes. The prepared powders were characterized by powder X-ray diffraction (PXRD) at NIIST (National Institute for Interdisciplinary Science and Technology, Council of Scientific and Industrial Research (CSIR)), Thiruvananthapuram, India, using an X-PERT PRO (Philips, Netherlands) X-ray diffractometer. The monochromator used for X-ray diffraction was copper which produces $CuK\alpha_1$ radiation as incident beam. The wavelength used for the X-ray intensity data collection was 1.54056 Å with a 2θ range of data collection from 10° to 120° with 0.05° step size.

$$2d \sin \theta = n\, \lambda$$

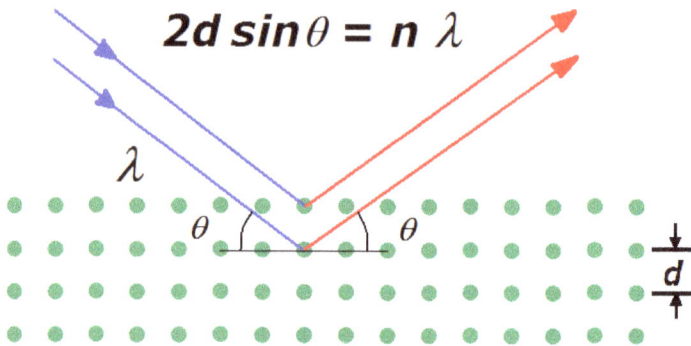

Figure 1.3 A view of X-ray Diffraction in accordance with Bragg's Law

1.9.2   Scanning electron microscope (SEM)

Scanning electron microscopy is an extremely useful tool to analyze the surface morphology of the samples which offers a better resolution than the optical microscope. The shape and morphology of particles are studied by scanning electron microscope (SEM). A schematic block diagram view is shown in figure 1.4. SEM uses a focused beam of high-energy electrons to generate a variety of signals at the surface of solid specimens. The beam of electrons is produced and accelerated from an electron source and passed through a series of condenser and objective lenses, which focus the electron beam. A scanning coil moves the beam across the specimen surface. Accelerated electrons carry significant amounts of kinetic energy and this energy is dissipated as a variety of signals produced by electron-sample interactions when the incident electrons are decelerated in the solid sample signals include secondary electrons (that produce SEM images), backscattered electrons (BSE), photons (characteristic X-rays that are used for elemental analysis), visible light (cathodoluminescence-CL) and secondary electrons and backscattered electrons are commonly used for imaging samples (Lawes, 1987).

Depending on the specimen and the equipment setup, the contrast in the final image provides information on the specimen composition, topography and morphology. In most applications, data are collected over a selected area of the surface of the sample, and a two dimensional image is generated that displays spatial variations in these properties. Areas ranging from approximately 1 cm to 5 microns in width can be imaged in a scanning mode using conventional SEM techniques (magnification ranging from 20x to approximately 30,000x, spatial resolution of 50 to 100 nm). Precise measurement of very small features and objects down to 50 nm in size is also accomplished using the SEM. In this study, a SHIMADZU SSX-550 Super scan SEM Model with EDS was used.

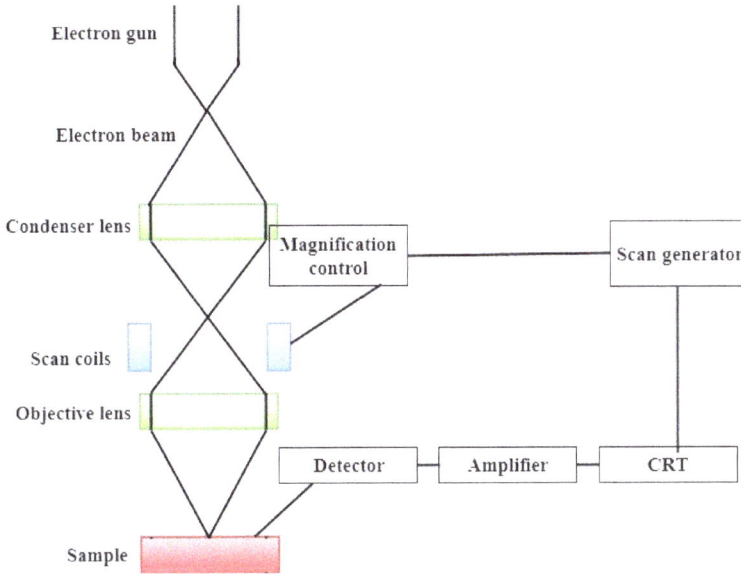

Figure 1.4. Schematic diagram of scanning electron microscope (SEM)

1.9.3 Energy dispersive X-ray spectroscopy (EDS)

The elemental composition of the synthesized powders in this study was determined by energy dispersive X-ray spectroscopy (EDS). EDS is a technique used for quantitative elemental analysis in conjunction with scanning electron microscopy (SEM). The EDS technique detects X-rays emitted from the sample during bombardment by an electron beam to characterize the elemental composition of the analyzed volume. Features or phases as small as 1 μm or less can be analyzed (Russ, 1984).

Information about the chemical composition of the sample is obtained by measuring the intensity distribution and energy of the signal generated by a focused electron beam impinging on the sample. The source of the electron is the electron gun of a scanning electron microscope. The incident beam of electrons interacts with core electrons of the sample's atoms transferring sufficient energy to it, thereby ejecting it from the target atom. This results in the creation of a hole within the atom's electronic structure. An electron from an outer, higher energy shell then occupies the hole releasing excess energy in the form of an X-ray photon. As a result of electronic transitions which occur between the outer and inner core levels a characteristic X-ray is emitted when the ionized atom 'relaxes' to a lower energy state by the transition of an outer-shell electron to the vacancy in the core shell which provide a quantitative and qualitative elemental composition of the sample (Russ, 1984). Due to a well-defined nature of the various atomic energy levels, it is clear that the energies and associated wavelengths of the set of X-rays will have characteristic values for each of the atomic species present in a sample as shown in figure 1.5.

A characteristic X-ray is usually emitted when the ionized atom 'relaxes' to a lower energy state by the transition of an outer-shell electron to the vacancy in the core shell. The X-ray is called characteristic because its energy equals the energy difference between the two levels involved in the transition and this difference is characteristic of the material. The EDS spectrum shows the frequency in counts of X-rays received for each energy level. The spectrum normally plots the peaks corresponding to the energy levels for which the most X-rays have been received. Each of these peaks corresponds to a specific atom, and therefore characteristic of a specific element. The intensity of a peak in the spectrum correlates with the concentration of the element in the sample.

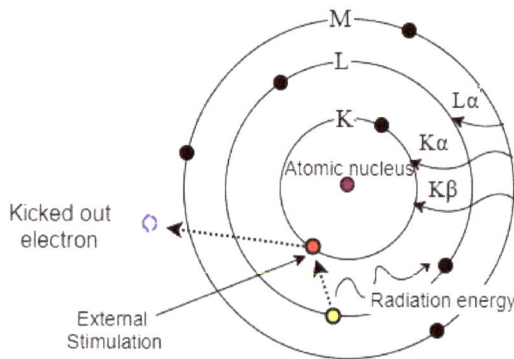

Figure 1.5. An inspection of the interaction of an electron beam with electrons within an atom

1.9.4 UV-visible spectroscopy

Ultraviolet-visible spectroscopy refers to absorption spectroscopy which ranges in the electromagnetic radiation between 190 nm to 800 nm which is classified into the ultraviolet and visible regions. In a standard UV-visible spectrophotometer, a beam of light is split; one half of the beam (the sample beam) is directed through a transparent cell containing a solution of the compound being analysed and one half (the reference beam) is directed through an identical cell that does not contain the compound but contains the solvent. The instrument is designed so that it can make a comparison of the intensities of the two beams as it scans over the desired region of the wavelengths. If the compound absorbs light at a particular wavelength, the intensity of the sample beam (IS) will be less than that of the reference beam (Gullapalli and Barron, 2010). Absorption of radiation by a sample is measured at various wavelengths and plotted by a recorder to give the spectrum which is a plot of the wavelength of the entire region versus the absorption (A) of light at each wavelength. And the band gap of the sample can be obtained by plotting the graph between (αhν vs hν) and extrapolating it along x-axis. Ultraviolet and visible spectrometry is almost entirely used for quantitative analysis; that is, the estimation of the amount of a compound known to be present in the sample. The sample is usually examined in solution. The block diagram of the UV-visible system is presented in figure 1.6.

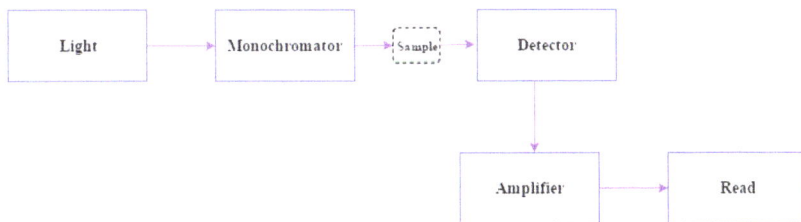

Figure 1.6. Block diagram of UV-visible spectrophotometer

1.9.5 Vibrating sample magnetometry

A vibrating sample magnetometer (VSM) is used to measure the magnetic parameters (like saturation magnetization, remanent magnetization, coercive field) of magnetic materials. It operates under the principle of Faraday's law (Simon Foner, 1959; Wesley Burgei, 2003) of electromagnetic induction

$$e = -N\frac{d}{dt}(BA\cos\theta) \qquad \dots\dots (1.28)$$

Where e is induced e.m.f. in the coil

- N is the number of turns in the coil
- B is the applied magnetic field
- A is the area of the coil
- θ is angle between applied magnetic field B and the direction normal to the surface of the coil.

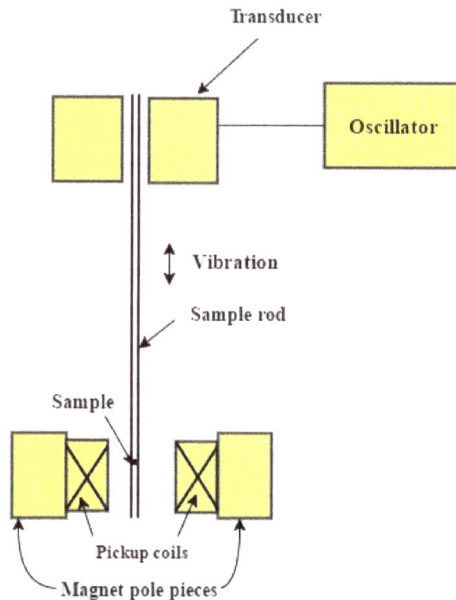

Figure 1.7. Schematic diagram of vibrating sample magnetometer (VSM)

araday's law states that a changing magnetic field will produce an electric field. In VSM, the magnetic sample to be studied is placed in a constant magnetic field. This magnetic field will magnetize the sample by aligning the magnetic spins with the field. If the applied magnetic field is stronger, then the magnetization will be larger. Depending upon this magnetic field and magnetization, magnetic materials are classified as ferromagnetic, ferrimagnetic, antiferromagnetic, paramagnetic and diamagnetic materials. These different magnetic behaviors of the materials can be studied through vibrating sample magnetometer (VSM) measurements as a function of magnetic field, time and temperature.

The schematic diagram of VSM is shown in figure 1.7. The essential parts of VSM are an electromagnet with power supply, vibration exciter, sample holder, magnetic sensor coils (pick-up coils), detector (Hall probe), amplifier and computer interface (Simon Foner, 1959). An electromagnet generates constant magnetic field which is used to magnetize sample.

1.10 Overview of the present work

1.10.1 $Zn_{1-x}Ti_xO$ (x= 0.01, 0.02, 0.03)

The substitution of Transition metal ions on Zinc oxide as host lattice may lead to some interesting behavioral changes in magnetic and optical properties. The present work discusses the effect of Ti dopant in Zinc oxide $Zn_{1-x}Ti_xO$ (x = 0.02 and x = 0.03) on the optical and magnetic properties. For that, Ti doped zinc-oxide (ZnO) was prepared by standard solid state reaction method in a high temperature furnace. The magnetic properties were analysed using vibrating sample magnetometry. The morphology of the samples was studied by SEM micrographs. The effect of the dopant on the semiconducting nature of the host ZnO was investigated by UV-vis studies

1.10.2 $Zn_{1-x}Fe_xO$ (x= 0.02, 0.04 and 0.06)

High purity Fe_2O_3 doped ZnO composites were grown by standard solid state reaction method in air at temperature range of 1350°C for 10 hours. The structures of the grown samples were investigated by X-ray diffraction studies. Single phase ZnO was observed indicating substitutional addition Fe^{2+} ion in Zn lattice. Electron density studies were carried out and sharing of the electrons between the atoms were observed in different planes and directions. Room temperature ferromagnetism can be observed in Fe-doped ZnO using vibrating sample magnetometry (VSM) with an increasing trend as the dopant concentration increases. Scanning electron microscopy (SEM) micrographs were taken to study of structure of the grown composites along with energy dispersive X-ray analysis (EDAX) to find the presence of Fe in ZnO. The energy gaps of the composites were measured by UV-vis spectra and they found to vary in range of 3.75- 3.80 eV.

1.10.3 $Zn_{1-x}V_xO$ (x= 0.02, 0.04 and 0.06)

This work investigates the changes in the structural and magnetic properties due to the doping of vanadium in zinc oxide (with varying concentration of vanadium, viz., 2%, 4%, and 6%) grown by solid state reaction method. The grown samples were analyzed through powder XRD measurements, vibrating sample magnetometry (VSM), field emission scanning electron microscopy (FESEM), energy dispersive X-ray analysis (EDAX) and

UV-vis spectroscopy. The electronic structure of grown $Zn_{1-x}V_xO$ has been studied through maximum entropy method (MEM) which reveals an increasing trend in the size of the Zn atom and also becoming more covalent. M-H curves obtained from the vibrating sample magnetometery (VSM) exhibit prominent room temperature ferromagnetism for x = 0.04. Field emission scanning electron microscopy (FESEM) and energy dispersive X-ray spectroscopy (EDAX) were employed for microstructure measurements.

1.10.4 $Zn_{1-x} Ni_{x/2}V_{x/2}O$ (x= 0.02, 0.04 and 0.06)

Powder samples of ZnO doped with Ni and V ($Zn_{1-x} Ni_{x/2}V_{x/2}O$) has been grown by standard solid state reaction technique with varying concentrations of x = 2%, 4 %, 6%. The grown samples were found to be dilute magnetic in nature along with semiconducting property. The structure of the material grown was solved using powder XRD refinements and the electronic structure of the material was determined by maximum entropy method (MEM). Magnetic properties of the samples were studied using vibrating sample magnetometry (VSM) and the samples exhibited ferromagnetic characteristics at room temperature. Atomic emission spectroscopy (AES) was employed to study the morphology of the samples. Direct band gap of the material was calculated using UV-vis studies.

References

[1] Bader R.F.W., Atoms in molecules, A Quantum Theory, Oxford University Press, (1991).

[2] Baomei W., Sader J.E., Boland J.J., Mechanical Properties of ZnO Nanowires, PRL 101, 175502 (2008). https://doi.org/10.1103/PhysRevLett.101.175502

[3] Bragg W.L., Proceedings of the Cambridge Philosophical Society, 17, 43 (1913).

[4] Catti M., Noel Y., Dovesi R., J. Phys. Chem. Solids. 64 (2003) 2183. https://doi.org/10.1016/S0022-3697(03)00219-1

[5] Collins D.M., Nature. 298, 49 (1982). https://doi.org/10.1038/298049a0

[6] Dietl T., Ohno H., Matsukura F., Cibert J., Ferrand D., Science, 287, 1019 (2000). https://doi.org/10.1126/science.287.5455.1019

[7] Gull S.F., Daniell G.J., Nature. 272, 686 (1978). https://doi.org/10.1038/272686a0

[8] Gullapalli S, Barron A. R, Characterization of Group 12-16 (II-VI) Semiconductor Nanoparticles by UV-visible Spectroscopy, OpenStax CNX, June, 2010 Online: Web site. http://cnx.org/content/m34601/1.1/.

[9] Iversen B.B., Larsen F.K., Figgis B.N., Reynolds P.A., Acta Cryst. B53, 923-932

Materials Research Forum LLC

https://doi.org/10.21741/9781644902257

(1996).

[10] Izumi F., Dilanian R.A., Recent Research Developments in Physics, Transworld Research Network, Trivandrum. Vol. 3, Part II, 699-726 (2002).

[11] Janisch R., Gopal P., Spaldin N.A., Journal of Physics: Condensed Matter, 17, R657-R689 (2005). https://doi.org/10.1088/0953-8984/17/27/R01

[12] Janotti A., Van de Walle C.G., Fundamentals of zinc oxide as a semiconductor, Reports on Progress in Physics, 72, 126501 (2009). https://doi.org/10.1088/0034-4885/72/12/126501

[13] Jaynes E.T., IEEE Trans. Syst. Sci. Cybern., SSC-4, 227 (1968). https://doi.org/10.1109/TSSC.1968.300117

[14] Jungwirth, T., Condensed Matter, Vol. 78, 802-830 (2006).

[15] King P.D.C., Veal T.D., J. Phys.: Condens. Matter 23, 334214 (2011). https://doi.org/10.1088/0953-8984/23/33/334214

[16] Klason P., Brseth T.M., Zhao Q.X., Svensson B.G., Kuznetsov A.Y., Bergman P.J., Willander M., Solid State Communication 145, 321-326 (2008). https://doi.org/10.1016/j.ssc.2007.10.036

[17] Kucheyev S.O., Bradby J.E., Williams J.S., Jagadish C., Swain M.V., Appl. Phys. Lett., 80, 956 (2002). https://doi.org/10.1063/1.1448175

[18] Lawes G., (1987) Scanning electron microscopy and X-ray microanalysis: Analytical chemistry by open learning, John Wiley & sons.

[19] Mallick P., Rath C., Biswal R., Mishra N.C., Ind. J. Phys., 83, 517 (2009). https://doi.org/10.1007/s12648-009-0012-4

[20] Malmros G., Thomas J.O., J. Appl. Cryst. 10, 7-11 (1977). https://doi.org/10.1107/S0021889877012680

[21] Meyer B.K., Alves H., Hofmann D.M., Kriegseis W., Forster D., Bertram F., Christen J., Hoffmann A., Straßburg M., Dworzak M., Haboeck U., Rodina A.V., phys. stat. sol. (b) 241, 231-260 (2004). https://doi.org/10.1002/pssb.200301962

[22] Mizokawa T., Nambu, T., Fujimori, A., Fukumura, T., Kawasaki, M., Physical Review B, 65, 085209 (2002). https://doi.org/10.1103/PhysRevB.65.085209

[23] Moe Brseth T., Svensson B.G., Kuznetsov A.Y., Klason P., Zhao Q.X., Willander M., Appl. Phys. Lett. 89, 262112-1-3 (2006). https://doi.org/10.1063/1.2424641

[24] Momma K., Izumi F., Commission on Crystallogr. Comput IUCr Newslett. 7, 106

(2006).

[25] Ohno, H., Making Nonmagnetic Semiconductors Ferromagnetic, Science, 281, 951-956 (1998). https://doi.org/10.1126/science.281.5379.951

[26] Pan Q., Huang K., Ni S., Yang F., Lin S., He D., J. Phys. D: Appl. Phys., 40, 6829 (2007). https://doi.org/10.1088/0022-3727/40/21/051

[27] Petříček V., Dušek M., Palatinus L., JANA 2006, The crystallographic computing system Institute of Physics Academy of sciences of the Czech republic, Praha (2006).

[28] Rao T.P., Kumar M.C.S., Ganesan V., Ind. J. Phys., 85, 1381 (2011). https://doi.org/10.1007/s12648-011-0184-6

[29] Rath C., Singh S. Mallick P., Pandey D., Lalla N.P., Mishra N.C., Ind. J. Phys., 83, 415 (2009). https://doi.org/10.1007/s12648-009-0018-y

[30] Rietveld H.M., J. Appl. Crystallogr., 2, 65 (1969). https://doi.org/10.1107/S0021889869006558

[31] Russ J.C., (1984) Fundamentals of Energy Dispersive X-ray Analysis, Butterworths, London.

[32] Sakata M., Sato M., Acta Cryst. A. 46, 263 (1990). https://doi.org/10.1107/S0108767389012377

[33] Simon Foner, Review of Scientific Instruments, 30, (7) (1959). https://doi.org/10.1063/1.1716679

[34] Sluiter M.H.F., Kawazoe Y., Sharma P., Inoue A., Raju A.R., Rout C., Waghmare U.V., Phys. Rev. Lett., 94, 187204 (2005). https://doi.org/10.1103/PhysRevLett.94.187204

[35] Stavola M., (1999), Identification of Defects in Semiconductors, Semiconductors and Semimetals, 51B, Academic press.

[36] Stout G.H., Jensen L.H., X-ray structure determination, chapter 1, 2nd Edition, Wiley- Interscience publication, (1989).

[37] Sundaresan A., Bhargavi R., Rangarajan N., Siddesh U., Rao C.N.R., Phy. Rev. B, 74, 161306(R) (2006). https://doi.org/10.1103/PhysRevB.74.161306

[38] Sundaresan A., Rao C.N.R., Nano Today, 4, 96 (2009). https://doi.org/10.1016/j.nantod.2008.10.002

[39] Wesley Burgei, Michael J. Pechan, and Herbert Jaeger, Am. J. Phys. 71, (8) (2003)

https://doi.org/10.1119/1.1572149

[40] Yu-Feng T., Shu-Jun H., Shi-Shen Y., Liang-Mo M., Oxide magnetic semiconductors: Materials, properties, and devices, Chinese Physics B, 22, 088505 (2013). https://doi.org/10.1088/1674-1056/22/8/088505

[41] Zhang Y.P., Yan S.S., Liu Y.H., Ren M.J., Fang Y., Chen Y.X., Liu G.L., Mei L.M., Liu J.P., Qiu J.H., Wang S.Y., Chen L.Y., Applied Physics. Letters, 89, 042501 (2006). https://doi.org/10.1063/1.2234280

[42] Zhao Q.X., Klason P., Willander M., Zhong H.M., Lu W., Yang J.H., Deep-level emissions influenced by O and Zn implantations in ZnO, Appl. Phys. Lett. 87, 211912-1-3 (2005). https://doi.org/10.1063/1.2135880

[43] http://wiki.verkata.com/en/wiki/Nanoparticle, Nanoparticle.

[44] http://www.aadet.com/article/nanoparticle, Nanoparticle Data.

[45] http://www.znoxide.org/properties.html Physical Properties of Zinc Oxide-CAS 1314-13-2 International Zinc Association.

[46] http://www.scribd.com/doc/43567667/Next-Generation-Technology-1

Chapter 2

Sample Preparation and X-Ray Diffraction Analysis

Abstract

The mechanical alloying method, developed by John Benjamin in the mid-1960 belongs to the solid state powder processing technique. The mechanical alloying method is based on the concept of synthesizing materials in a non-equilibrium state by "energizing and quenching". Energizing brings the material into a highly non-equilibrium (metastable) state by some external dynamical force and quenching brings the material into a configurationally frozen state which will be used as a precursor to obtain final desired sample by subsequent sintering at high temperatures. Mechanically alloyed synthesized materials possess improved physical and mechanical characters in comparison with conventional processed materials. The samples synthesized by mechanical alloying are usually subjected to a sintering process. Sintering controls the extrinsic properties such as particle size, shape, distribution, porosity, and agglomeration etc., of the samples. The sintering process is a two stage process (a) pre-sintering and (b) final sintering. The as prepared samples are mixed thoroughly by grinding and then pressed into pellets using a hydraulic press followed by heat treatment for some specific duration. The pre-sintered pellets were grounded into fine powder so as to reduce the particle size and to promote the mixing of any un-reacted oxides. After grinding, the powder is then pressed into the required shape by applying a pressure using a hydraulic press. The pellets/torroids prepared in this manner are sintered at the desired temperature for some specific duration. The final sintering increases the density and decreases the porosity of the material. The oxide based dilute magnetic semiconductors were synthesized in the present work were characterized for atructural and average particle size analysis using X-ray diffraction studies.

Keywords

Sintering, High Temperature Furnace, Solid State Reaction (SSR), DMS Material, X-Ray Diffraction

2.1 Introduction

In this work, four different DMS materials have been synthesized by solid state reaction method. The synthesized DMS materials are as follows:

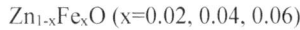

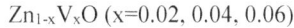

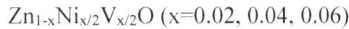

$$Zn_{1-x}Ti_xO \ (x=0.02, \ 0.03)$$

$$Zn_{1-x}Fe_xO \ (x=0.02, \ 0.04, \ 0.06)$$

$$Zn_{1-x}V_xO \ (x=0.02, \ 0.04, \ 0.06)$$

$$Zn_{1-x}Ni_{x/2}V_{x/2}O \ (x=0.02, \ 0.04, \ 0.06)$$

In this chapter the sample preparation procedures and the results obtained from powder X-ray characterization and scanning electron microscopy have been presented. The synthesized DMS materials have been stucturally characterized by the powder X-ray diffraction method (PXRD). Table 2.1 shows the details of molecular weight and purity of the samples used. The experimental X-ray diffraction data sets have been refined using Rietveld method (Rietveld, 1969) through JANA 2006 (Petříček *et al.*, 2006) software. The Rietveld refinement technique (Rietveld, 1969) refines various parameters such as lattice parameters, fractional coordinates, *etc.*

Table 2.1. Molecular weights and purity of chemicals used

Chemicals	Molecular weight (gm/mol)	Purity (%)
Zn O	100.05	99.99
V_2O_5	20.01	99.99
Ni O	50.01	99.97
Ti O_2	20.00	99.98
Fe_2O_3	20.00	99.95

2.2 Solid state reaction (SSR)

This is the most widely used method for the synthesis of polycrystalline bulk materials. Solid state reaction (SSR) method provides a large range of selection of starting materials like, oxides, carbonates, etc. Since solids do not react with each other at room temperature (RT), it is necessary to heat them at elevated temperatures for the proper reaction to take place at appreciable rate. In SSR method, the solid reactants react chemically without the presence of any solvent at high temperatures yielding a product which is stable. The major advantage of SSR method is the final product in solid form is structurally pure with the desired properties depending on the final sintering temperatures. This method is environmentally friendly and no toxic or unwanted waste is produced after the SSR is complete. Figure 2.1 shows various steps involved in conventional SSR route. The final product of SSR is usually in the form of a powder or a sintered, polycrystalline piece. Large single crystals are not usually obtained by this method.

Take appropriate high purity starting materials, fine grain powders, in stoichiometric proportions.

- Weigh the starting materials, as per the stoichiometric ratios.
- Mix them together, thoroughly using agate mortar and pestle or ball milling (for large quantity sample).
- Heat the solid powder mixture (calcination) at elevated temperatures in air using a muffle furnace.
- Repeat the calcination process twice with intermittent grinding.
- The black powder is pelletized to form uniform and compact pellets which are sintered at more elevated temperatures ($\sim1375^\circ$C) for prolonged duration.

The final temperature and duration of sintering may vary depending on the nature and properties of the sample under preparation. The solid state reaction is the most widely used method for the preparation of polycrystalline solids from a mixture of solid starting materials. Solids of starting material can't react at room temperature and it is necessary for the reaction to occur at an appreciable rate. Hence the starting materials are heated to higher temperatures of about 1000°C to 1500°C. The solid state reactions depend on several factors, viz., reaction conditions, structural properties of the reactants, surface area of the solids, their reactivity and the thermodynamic free energy change associated with the reaction (Narayan *et al.*, 2009).

Solid State Reaction

Stoichiometric   Quantities

Starting Material   Starting Material   Starting Material   Starting Material

Dry Mixing in Powder   Form in Agate Mortar

1st Heating: Calcination: 900°C /24 hrs

Grounding Thoroughly   and Pelletization

2nd Heating: Sintering: 1100°C /48 hrs

Grounding Thoroughly   and Pelletization

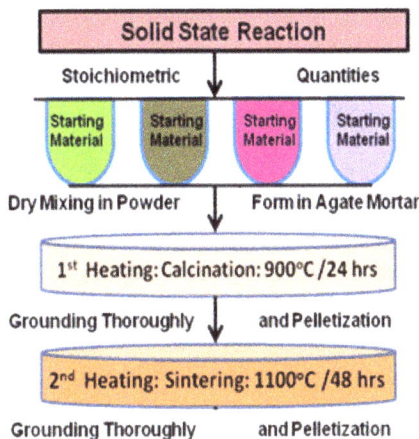

Figure 2.1 Schematic illustration of sample preparation by conventional solid state reaction route

2.2.1 Advantages of solid state reaction

- It limits the formation of side products.

- Solvents are not necessary in the reaction

- Products do not require extensive purification to remove traces of solvent and impurities

- These reactions will take just a few minutes rather than hours to complete because the reactants are in intimate contact with each other.

- There is the possibility to form organometallic complexes in the solid state (Braga et al., 2002; Desiraju et al., 1989).

- Even the Cascade reactions can be carried out quite easily and with high yields (Bradly, 2002).

Reaction between or reaction within the solid materials without any solvent (i.e.) solvent less is called Solid state reaction. Solid-state chemical reaction usually occurs at the boundaries of particles where phases are in contact. Whereas the process is proportional to temperature, as temperature increases, the rate of diffusion increases (Callister et al., 2007).

In this reaction, atoms/ions have to interact, in order to interact atoms/ions have to diffuse. Temperature has the greatest influence in involving the atomic diffusion process. As the reaction proceeds the compound forms at the particle boundaries creating a barrier to diffuse. This decreases the rate of reaction, for the reactants having to diffuse across larger distances. In order to address this problem, the reactant mixture is reground and re-pelletized. To diffuse an atom through a solid two conditions are needed. There must be a neighboring site that is empty. The atom has to have enough energy to break its bonds and then cause some lattice distortion. Diffusion can occur in two different ways, through vacancies or through interstitial sites. Vacancy diffusion will be looked at for this reaction since the atoms that are being reacted are too large to diffuse through interstitial sites.

In our work, the raw materials are weighed using the electronic balance of the model MK 200E which has the readability accuracy of 0.001 gm. For mixing and grinding, an agate mortar and pestle were initially used. For pelletizing the samples, a pelletizer which uses a maximum hydraulic pressure of 100 MPa has been used. Two different dies of 1.0 cm and 1.2 cm diameter were used to get the pellets of the prepared materials. For calcinations and sintering, a tubular furnace was used. This furnace has a working temperature up to 1600°C. This tubular furnace has a programmable heating rate of 1°C/min to 5°C/min and has 1°C accuracy of dwell temperature.

2.3 Synthesis of DMS materials

2.3.1 Synthesis of $Zn_{1-x}Ti_xO_2$

The $Zn_{1-x}Ti_xO$ powders with varying concentrations as x=0.02, 0.03 were prepared using the solid state reaction technique. High purity TiO_2 (Alfa aesar, 99.9%) and ZnO (Alfa aesar, 99.99%) were mixed in a molar ratio are ground for 1 h to obtain homogenous mixture. The samples with different concentrations were pelletized and kept in an alumina crucible separately. The samples were sintered at 1350°C for 10 h in high temperature furnace. The sintered samples were taken and ground into fine powders.

2.3.2 Synthesis of $Zn_{1-x}Fe_xO$

The system $Zn_{1-x}Fe_xO$, (x=0.02, 0.04, 0.06) were prepared by mixing stochiometric quantities of pure powders of zinc oxide (Alfa aesar, 99.99%) and Fe_2O_3 (Alfa aesar, 99.95%). The mixed powders were ground well using an agate mortar and pestle. The systems with various doping concentrations were prepared by standard solid state reaction method. Initially, the mixed powders were compressed to form dense pellets and calcined at a temperature of 1350°C for 10 h at a rate of 5°C/min in air. Then the furnace was room cooled. The samples were taken out and ground well as smooth powders.

2.3.3   Synthesis of $Zn_{1-x}V_xO$

The system $Zn_{1-x}V_xO$ (x=0.02, 0.04, 0.06) was prepared by mixing stochiometric quantities of pure powders of zinc oxide (Alfa aesar, 99.99%) and vanadium pentoxide (Alfa aesar, 99.95%). The mixed powders were ground well using an agate mortar and pestle. The system with various doping concentrations, was prepared by the standard solid state reaction method. Initially, the mixed powders were compressed to form dense pellets and calcined at a temperature of 800°C for 10 h at a rate of 5°C/min in air. Then, the furnace was room cooled and the calcined pellets were again powdered and compressed to form dense pellets. These pellets were calcined again at a high temperature of 1100°C for 15 hours with the heating rate of 5 °C/min and then the furnace was room cooled. The samples were taken out and ground well into smooth powders.

2.3.4   Synthesis of $Zn_{1-x}Ni_{x/2}V_{x/2}O$

Ni and V co-doped ZnO was prepared by the standard solid state reaction method. Nickel powder (Alfa aesar, 99.99%) and vanadium pentoxide (Alfa aesar, 99.99%) were mixed thoroughly and grounded in zinc oxide powder (Alfa aesar, 99.99%) under desired stochiometric compositions such as x=0.02, 0.04, 0.06. The mixed powders were grounded nicely in agate and mortar pestle for 1 h individually. The well ground samples were separated according to their compositions and pelletized in a 5 ton hydraulic pelletizer. The pellets were taken in alumina crucible and sintered in high temperature furnace for 8 hours at a temperature of 950°C. The furnace was then room cooled and the samples were ground into smooth powders. Powder x-ray diffraction (XRD) data (CuKα) was collected to check the structure and phase purity of the samples.

2.4   X-ray diffraction analysis

2.4.1 $Zn_{1-x}Ti_xO$ (x=0.01, 0.02, 0.03)

The raw powder X-ray diffraction patterns of $Zn_{1-x}Ti_xO$ at concentration levels x=0.01, x=0.02 and x=0.03 are shown in figure 2.1 (a). Figure 2.1 (b) shows enlarged view of XRD patterns of the samples for the nominal concentrations at x=0.01, x=0.02 and 0.02. The prominent XRD peaks show host peaks of ZnO (JCPDS # 361451) which confirms the host lattice as ZnO with space group P63mc. The presence of intense additional phases at Bragg reflections (220) and (311) shown in figure 2.1(b) corresponds to Zn_2TiO_4 (JCPDS # 130536) with space group Fd-3m. Figures 2.1 (c) - (h) show shifting of all the Bragg peaks towards lower values of diffraction angles with increase in Ti^{4+} concentration indicating increase in the lattice parameter. This is due to the ionic radius of Ti^{4+} (0.68 Å) (Singh et al., 2008) is smaller than that of Zn^{2+} (0.74 Å) (Singh et al., 2008), and thus Ti^{4+} ions can replace Zn^{2+} ions at substitutional sites. The existence of zinc titanate additional phase was reported by Stanley et al., (Stanley et al., 2011). The presence of an additional phase was identified to be zinc orthotitanate (Zn_2TiO_4). Zn_2TiO_4 has a face-centered cubic crystal structure with a lattice parameter a = 0.8460(2) nm.

The Bragg peaks show a left shift from the zinc oxide with increase in concentration (Table 2.2). This peak position shifting clearly indicates that Ti^{4+} ion has incorporated into the ZnO host lattice (Chikoidze et al., 2005). If addition of Ti^{4+} ion takes place in the Zn^{2+} lattice, eventually the lattice parameters is expected to get enhanced (Table 2.2) due to inclusion of Ti^{4+} ion as reported elsewhere (Chikoidze et al., 2005). In the present study, the cell parameters were refined using unit cell refinement software (Holland et al., 1997) and the results obtained are shown in Table 2.3.

2.4.2 Rietveld analysis

Rietveld refinement (Rietveld, 1967) is the measuring means which was devised by Hugo Rietveld (Rietveld, 1967). It is used for the characterization of crystalline materials. The raw intensities were refined using the software program JANA 2006 (Petříček, 2006). The reliability indices factors from the refinement were shown in table 2.4. This technique reduces the differences between calculated profile and observed profile. The fitted profiles coincides well with observed and calculated intensities which are shown in the figure 2.1.(i), 2.1(j) and figure 2.1 (k) for x = 0.01, x = 0.02 and x = 0.03compositions. The composition x = 0.01 did not show any significant variations with its host ZnO because of its low concentration nature.

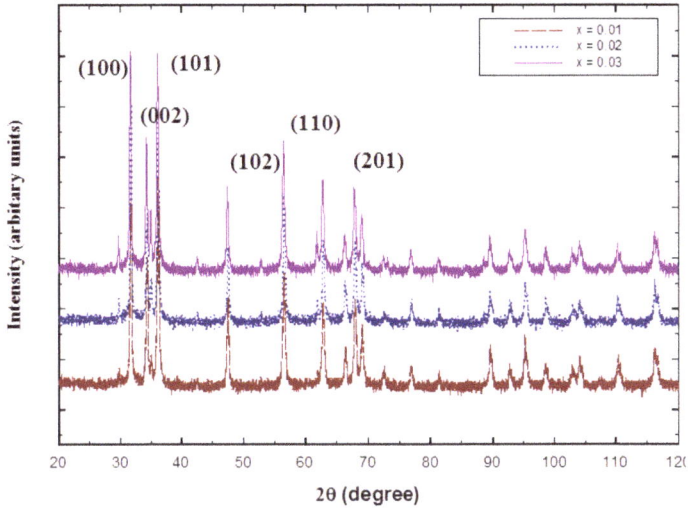

Figure 2.1 (a) XRD patterns of $Zn_{1-x}Ti_xO$ (x=0.01, 0.02, 0.03) samples

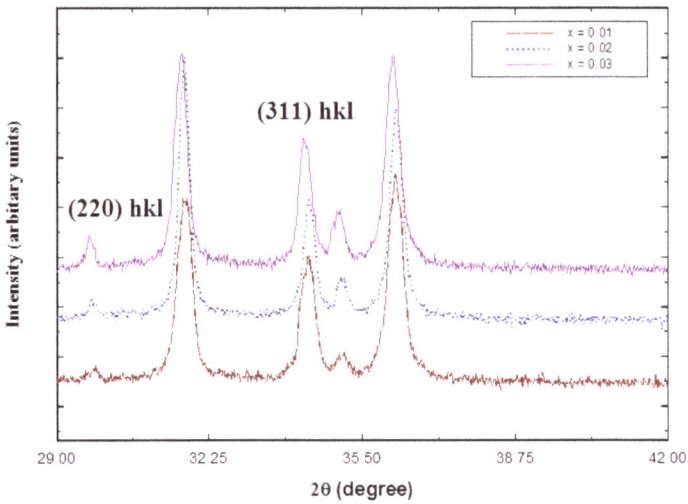

Figure 2.1 (b) Enlarged view of (220) and (311) hkl planes

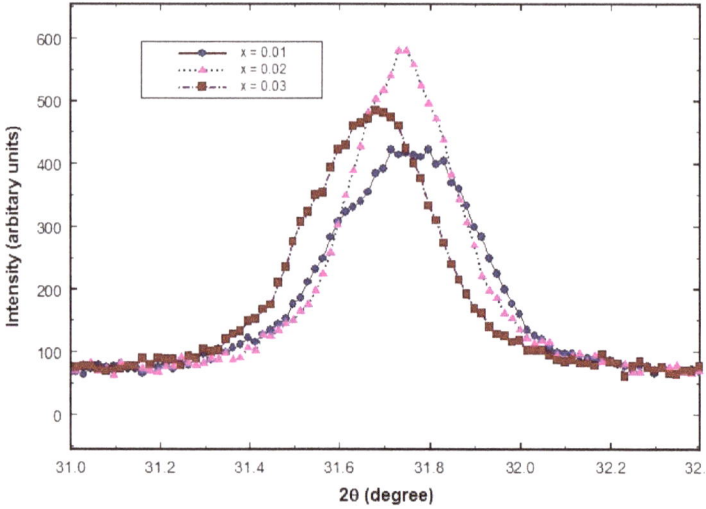

Figure 2.1 (c) Bragg peak shift on (100) plane

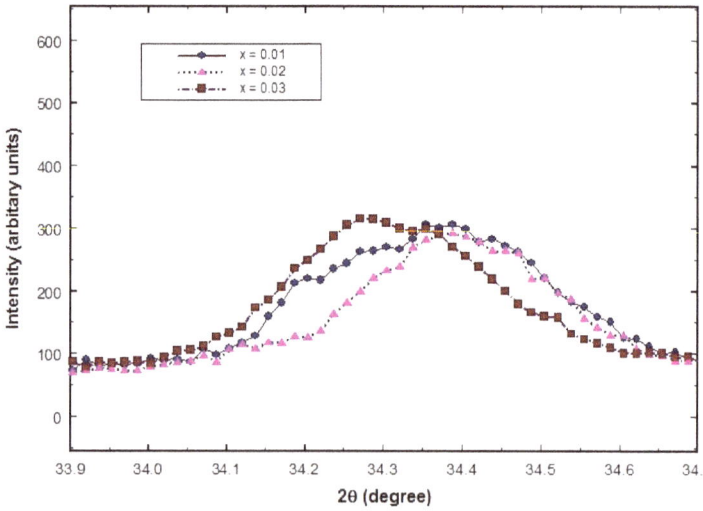

Figure 2.1 (d) Bragg peak shift on (002) plane

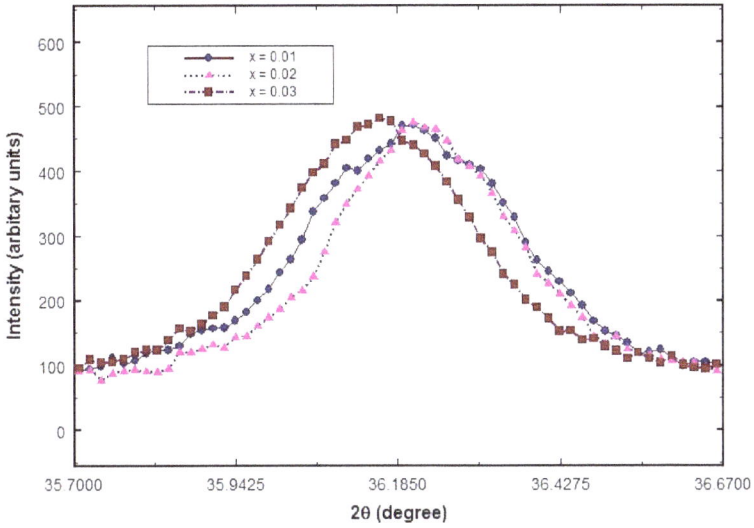

Figure 2.1 (e) Bragg peak shift on (101) plane

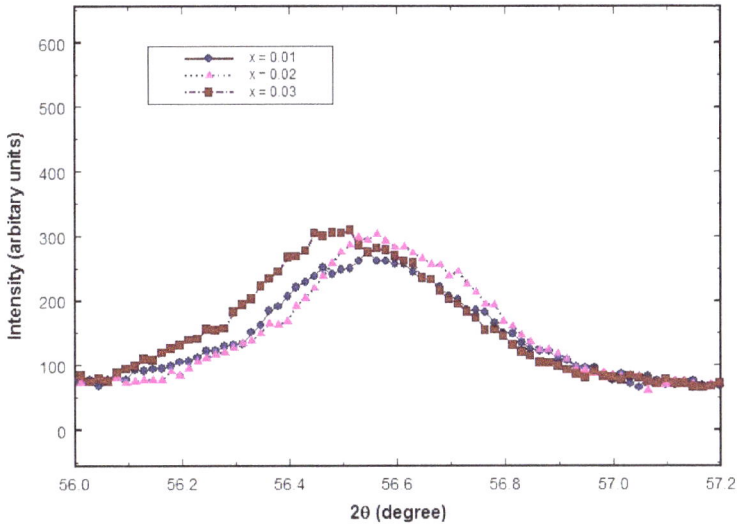

Figure 2.1 (f) Bragg peak shift on (110) plane

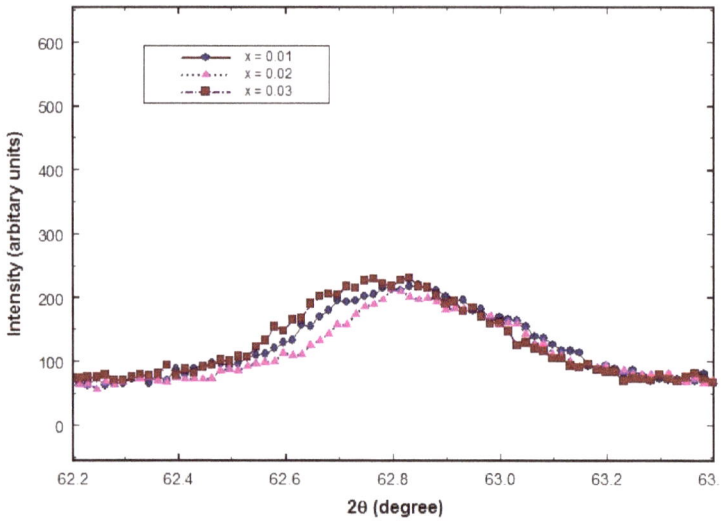

Figure 2.1 (g) Bragg peak shift on (103) plane

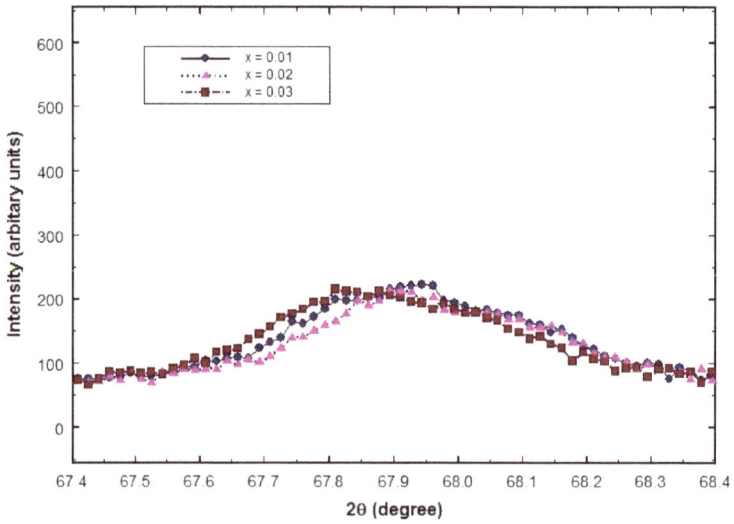

Figure 2.1 (h) Bragg peak shift on (201) plane

Materials Research Forum LLC
https://doi.org/10.21741/9781644902257

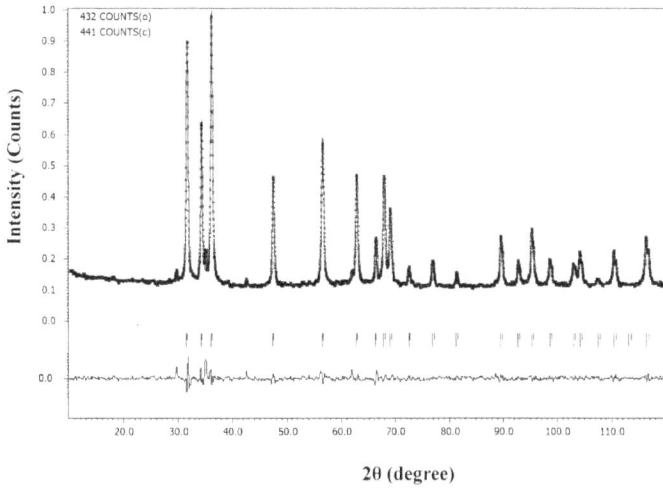

Figure 2.1 (i) Rietveld refined powder XRD profile for $Zn_{0.99}Ti_{0.01}O$

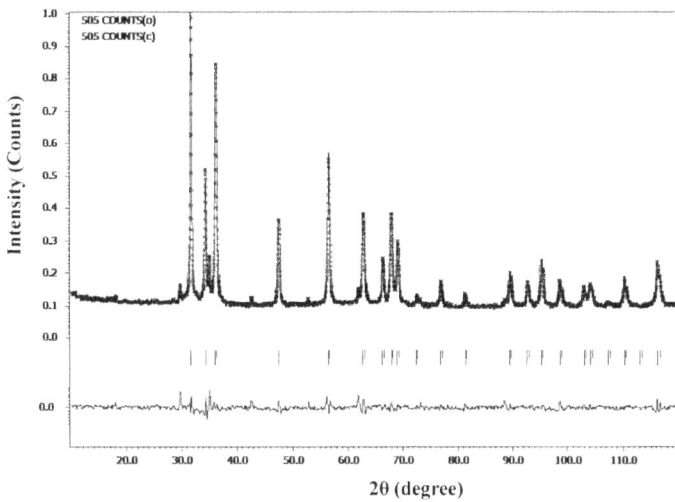

Figure 2.1 (j) Rietveld refined powder XRD profile for $Zn_{0.98}Ti_{0.02}O$

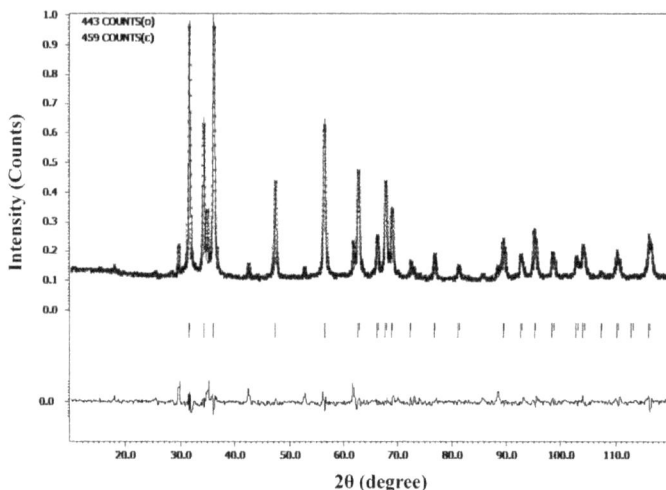

Figure 2.1 (k) Rietveld refined powder XRD profile for $Zn_{0.97}Ti_{0.03}O$

Table 2.2 Variation in Bragg peak shifts in the observed XRD spectra for $Zn_{1-x}Ti_xO$

(hkl)	(100)		(002)		(101)	
System	Peak distance (2θ)	Intensity	Peak distance (2θ)	Intensity	Peak distance (2θ)	Intensity
x=0.01	31.77	57	34.42	44	36.25	100
x=0.02	31.75	598	34.39	299	36.19	497
x=0.03	31.67	496	34.27	304	36.15	499

Table 2.3 Structural parameters for $Zn_{1-x}Ti_xO$ from unit cell refinement

Parameter	x = 0.01	x = 0.02	x = 0.03
a (Å)	2.2502	2.2508	2.2541
c (Å)	5.2098	5.2133	5.2167

Table 2.4. Reliability indices for the sample $Zn_{1-x}Ti_xO$ for various concentrations

Composition	Refinement parameters	x=0.01	x=0.02	x=0.03
$Zn_{1-x}Ti_xO$	$R_{obs}(\%)$	8.67	6.53	5.51
	$wR_{obs}(\%)$	8.59	8.09	7.09
	$R_p(\%)$	11.55	8.76	8.94
	$wR_p(\%)$	15.09	11.38	11.77
	GOF	1.23	0.92	0.97

2.4.3 $Zn_{1-x}Fe_xO$ (x=0.02, 0.04, 0.06)

Powder X-ray diffraction datasets were collected for the powder samples, using monochromatic incident beam of CuKα (1.54056Å) radiation using Philips X-perto diffractometer, with 2θ ranging from 10° to 120° under common X-ray spectrometer settings. Figure 2.2 (a) shows XRD patterns for three compositions x=0.02, 0.04 and 0.06. The atomic number of zinc atom is 30 which are greater than the atomic number of Ferrous which is 26, which reveals itself as shown in these figures that as the concentration of Fe increases, the intensity of XRD peaks decreases. Figures 2.2 (b)-(g) represent different Bragg peaks ranging from 20° to 60° for the three compositions respectively. The ionic radius of Fe^{2+} (0.77 Å) (Wu et al., 2006) ion is smaller than the Zn^{2+} (0.88 Å) ion (Wu et al., 2006). The Bragg peaks show a right shift from the zinc oxide with increase in concentration (Table 2.5). If addition of Fe^{2+} ion takes place in the Zn^{2+} lattice, eventually the lattice parameters are expected to get enhanced (Table 2.6) due to inclusion of Fe^{2+} ion as reported elsewhere (Chikoidze et al., 2005). In the present study, the cell parameters were refined using unit cell refinement software (Holland et al., 1997) and the results obtained are shown in Table 2.6.

The observed raw X-ray profiles of $Zn_{1-x}Fe_xO$ with different compositions x=0.02, 0.04 and 0.06 are used in the Rietveld method (Rietveld, 1969) which employs in the software JANA 2006 (Petříček et al., 2006) by considering the cubic phase with space group $Fm\bar{3}m$. The fitted profiles for XRD data at x=0.02, 0.04, 0.06 are shown in figures 2.2 (h)-(j) respectively. The values of reliability indices were depicted in table 2.7.

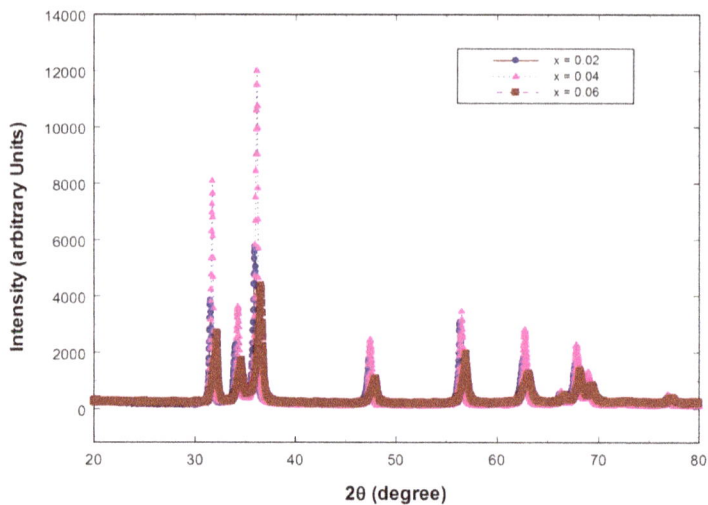

Figure 2.2 (a) XRD patterns of $Zn_{1-x}Fe_xO$ (x=0.02, 0.04, 0.06) samples

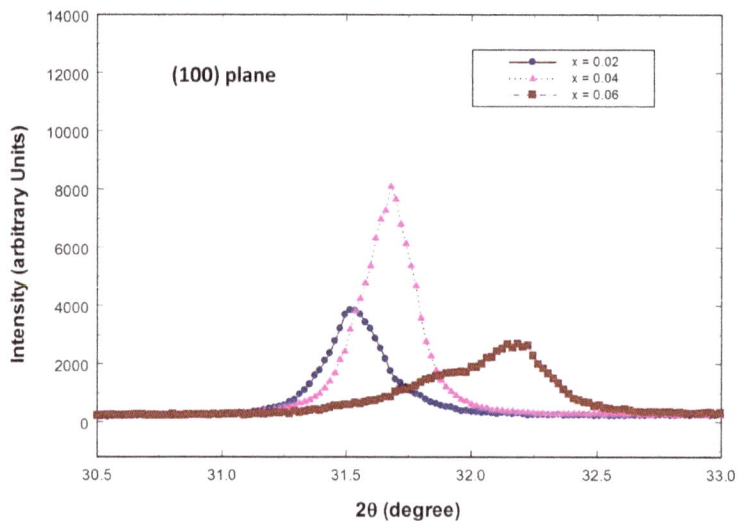

Figure 2.2 (b) Observed X-ray powder diffractograms of $Zn_{1-x}Fe_xO$ (x=0.02, 0.04, 0.06) at $2\theta=30.5°\text{-}32.0°$

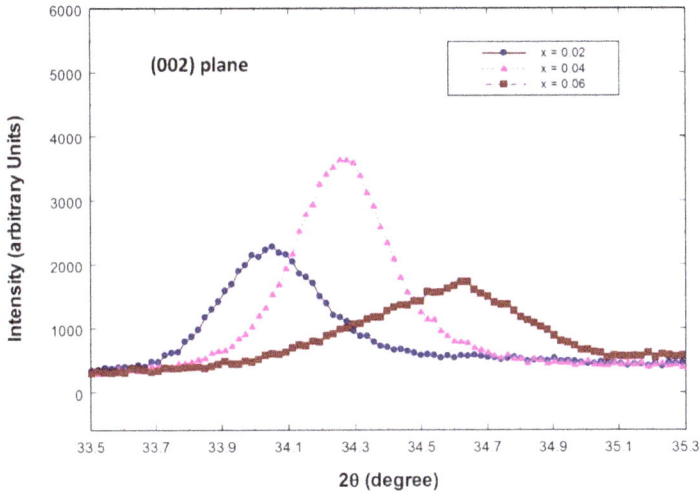

Figure 2.2 (c) Observed X-ray powder diffractograms of $Zn_{1-x}Fe_xO$ (x=0.02, 0.04, 0.06) at $2\theta=34°-35.4°$

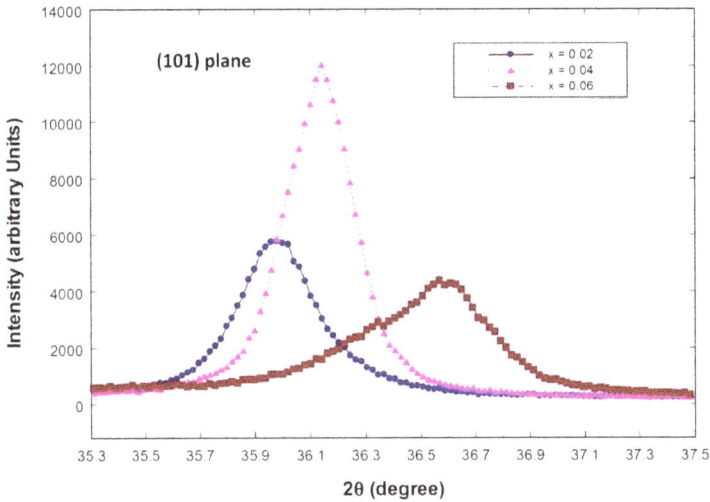

Figure 2.2 (d) Observed X-ray powder diffractograms of $Zn_{1-x}Fe_xO$ (x=0.02, 0.04, 0.06) at $2\theta=36°-37°$

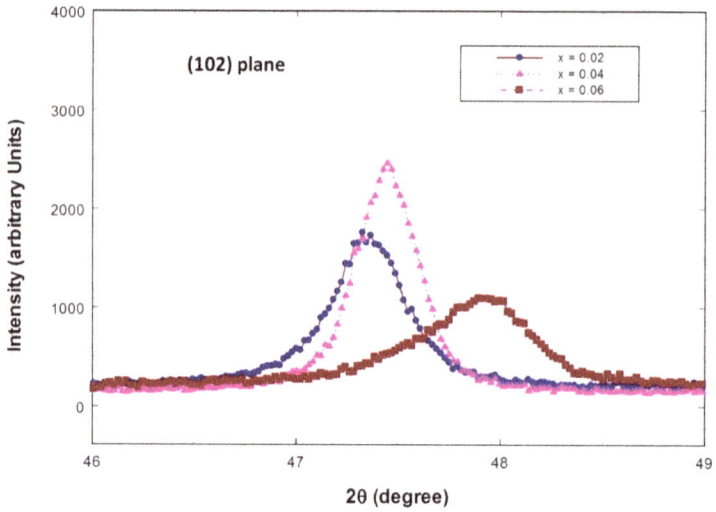

Figure 2.2 (e) Observed X-ray powder diffractograms of $Zn_{1-x}Fe_xO$ (x=0.02, 0.04, 0.06) at $2\theta=47.2°-48.4°$

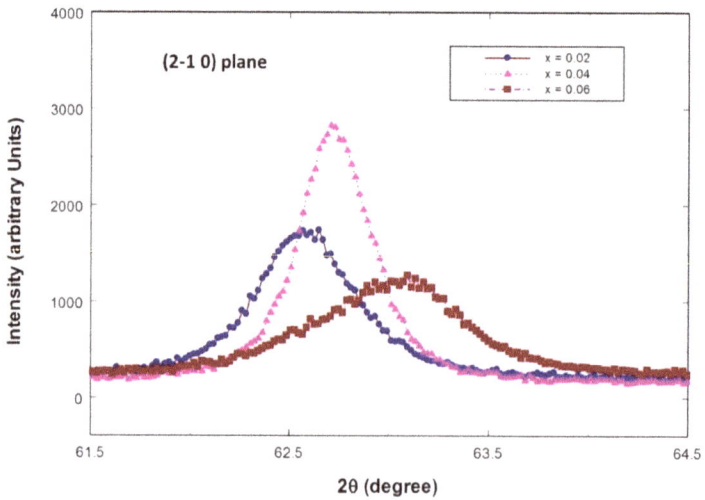

Figure 2.2 (f) Observed X-ray powder diffractograms of $Zn_{1-x}Fe_xO$ (x=0.02, 0.04, 0.06) at $2\theta=56.3°-57.6°$

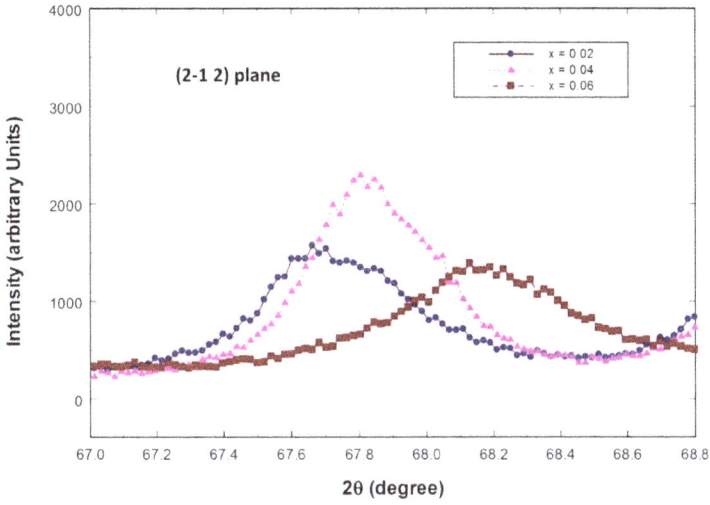

*Figure 2.2 (g) Observed X-ray powder diffractograms of $Zn_{1-x}Fe_xO$ (x=0.02, 0.04, 0.06)
at $2\theta=62.6°-62.6°$*

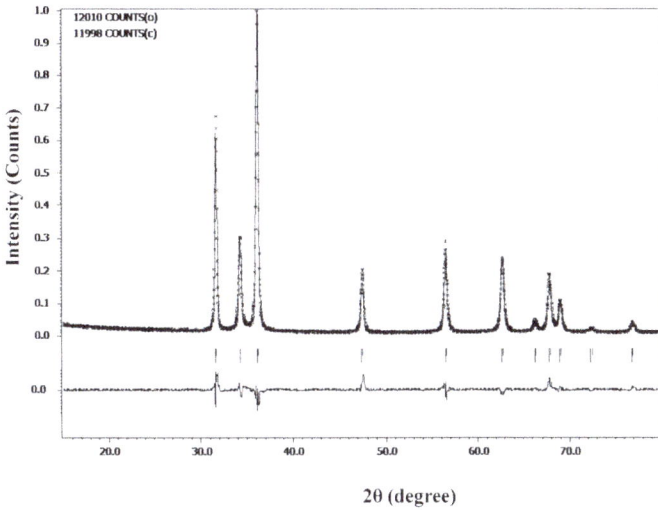

Figure 2.2 (h) Rietveld refined powder XRD profile for $Zn_{0.98}Fe_{0.02}O$

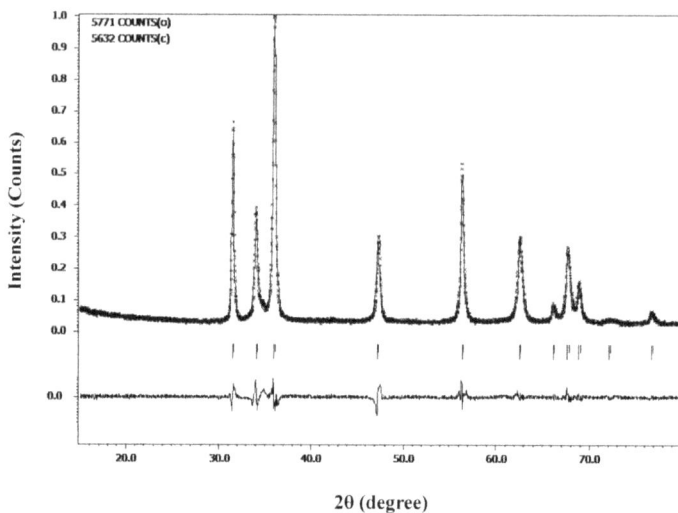

Figure 2.2 (i) Rietveld refined powder XRD profile for $Zn_{0.96}Fe_{0.04}O$

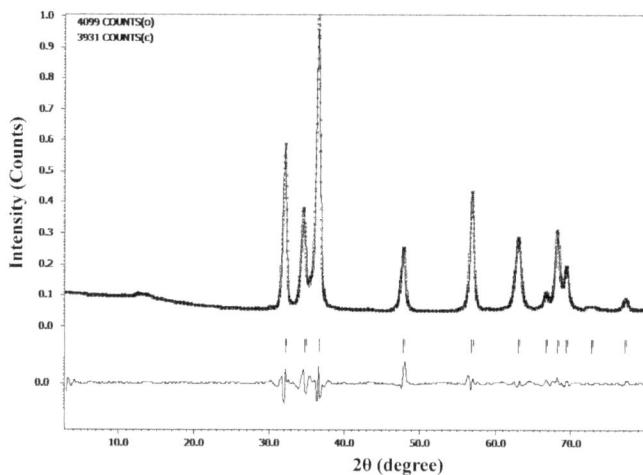

Figure 2.2 (j) Rietveld refined powder XRD profile for $Zn_{0.94}Fe_{0.06}O$

Table 2.5 XRD Bragg peak shifting of $Zn_{1-x}Fe_xO$

(hkl) plane	x=0.02	x=0.04	x=0.06	ZnO (*JCPDS)
(100)	31.51	31.68	32.18	31.77
(002)	32.97	34.23	34.62	34.42
(101)	35.95	36.23	36.58	36.25
(102)	47.38	47.46	47.85	47.53
(103)	62.51	62.79	62.18	62.86

*JCPDS reference number: 361451

Table 2.6 Structural parameters for $Zn_{1-x}Fe_xO$ from unit cell refinement

Parameter	x=0.02	x=0.04	x=0.06
a (Å)	2.2259	2.2386	2.2498
c (Å)	5.1873	5.2188	5.2612

Table 2.7 Reliability indices for $Zn_{1-x}Fe_xO$ for various concentrations

		x=0.02	x=0.04	x=0.06
$Zn_{1-x}Fe_xO$	$R_{obs}(\%)$	2.55	0.91	1.56
	$wR_{obs}(\%)$	2.02	1.11	1.76
	$R_p(\%)$	6.18	6.63	6.42
	$wR_p(\%)$	8.13	9.05	8.29
	GOF	1.64	1.71	1.64

2.4.4 $Zn_{1-x}V_xO$ (x=0.02, 0.04, 0.06)

Figure 2.3 (a) shows powder X-ray diffraction patterns of $Zn_{1-x}V_xO$ (x=0.02, 0.04, 0.06) samples. Figures 2.3 (b)-(g) represent different Bragg peaks ranging from 20° to 60° for the three compositions respectively. The ionic radius of V^{2+} (0.93Å) ion is larger than the Zn^{2+} (0.60 Å) ion (Karamat et al., 2010). The atomic number of zinc atom is 30 which is greater than the atomic number of vanadium which is 23, which reveals as shown in these figures that as the concentration of vanadium increases, the intensity of XRD peaks decreases. The Bragg peaks show a left shift from the zinc oxide, (JCPDS reference no: 361451) with increase in concentration (Table 2.8). This peak position shifting clearly indicates that V^{2+} ion has incorporated into the ZnO host lattice. These figures clearly indicate the intensity of Bragg peak decreases from 2% to 6% of vanadium doping. If substitution of V^{2+} ion takes place in the Zn^{2+} lattice, eventually the lattice parameters are expected to get enhanced due to larger ionic radius of V^{2+} ion as reported elsewhere (Karamat et al., 2010). In the present study, the structural parameters were refined using unit cell refinement technique, which is more widely used to study the characterization of crystalline materials. The results obtained are shown in Table 2.9. In mixed solution, the distorted lattice constant is expected to lie between the lattice constants of the pure elements. If the dependence is linear, then the solution will obey Vegard's law (Karamat et al., 2010). The lattice constants obtained by Vegard's law for $Zn_{1-x}V_xO$ are plotted in figure 2.4 perfectly doped in all the three different compositions, which can be seen in fitted profiles, which clearly show the presence ZnO peaks without any additional phases of V_2O_5.

2.4.4.1 Rietveld analysis

The observed raw X-ray profiles of $Zn_{1-x}V_xO$ with different compositions x=0.02, 0.04 and 0.06 are used in the Rietveld method (Rietveld, 1969) which employs in the software JANA 2006 (Petříček et al., 2006) by considering the cubic phase with space group $Fm\overline{3}m$. The fitted profiles for XRD data at x=0.02, 0.04, 0.06 are shown in figures 2.5 (a)-(c) respectively. The refined structural values are tabulated in Table 2.10. Table 2.10 gives the observed and calculated structure factors of grown $Zn_{1-x}V_xO$ for three different compositions of x=0.02, 0.04, 0.06 respectively. From the values of reliability indices in Table 2.10, we observed that the grown $Zn_{1-x}V_xO$ has been perfectly fitted through the software JANA2006 (Petříček et al., 2006).

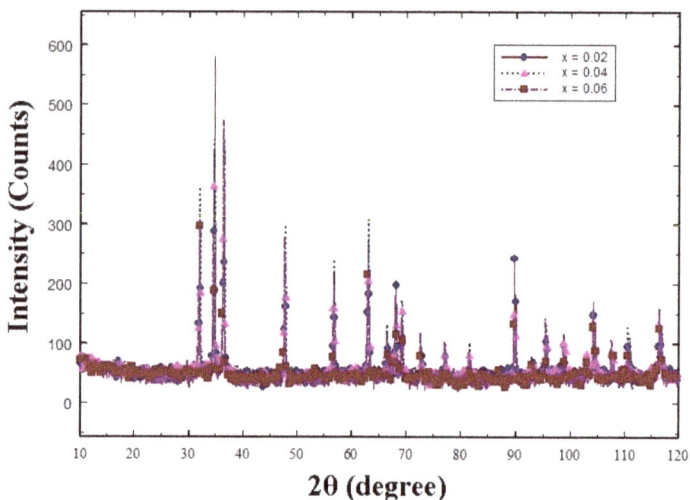

Figure 2.3 (a) XRD patterns of $Zn_{1-x}V_xO$ (x=0.02, 0.04, 0.06) samples

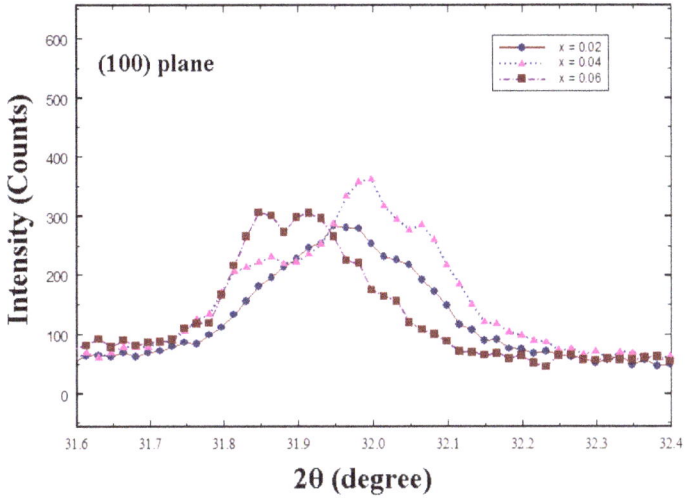

Figure 2.3 (b) Observed X-ray powder diffractograms of $Zn_{1-x}V_xO$ (x=0.02, 0.04, 0.06) at $2\theta=31.6°-32.4°$

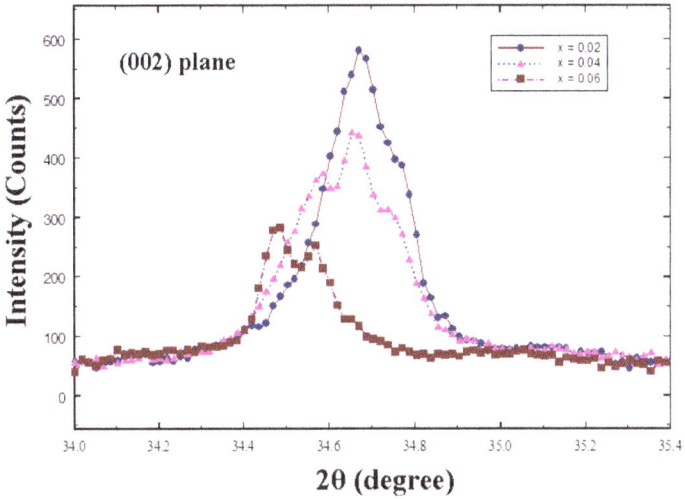

Figure 2.3 (c) Observed X-ray powder diffractograms of $Zn_{1-x}V_xO$ (x=0.02, 0.04, 0.06) at $2\theta=34°-35.4°$

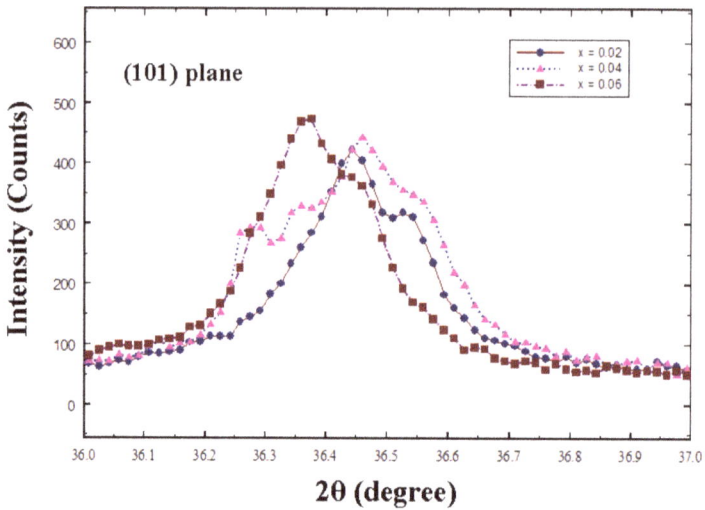

Figure 2.3 (d) Observed X-ray powder diffractograms of $Zn_{1-x}V_xO$ (x=0.02, 0.04, 0.06) at $2\theta=36°-37°$

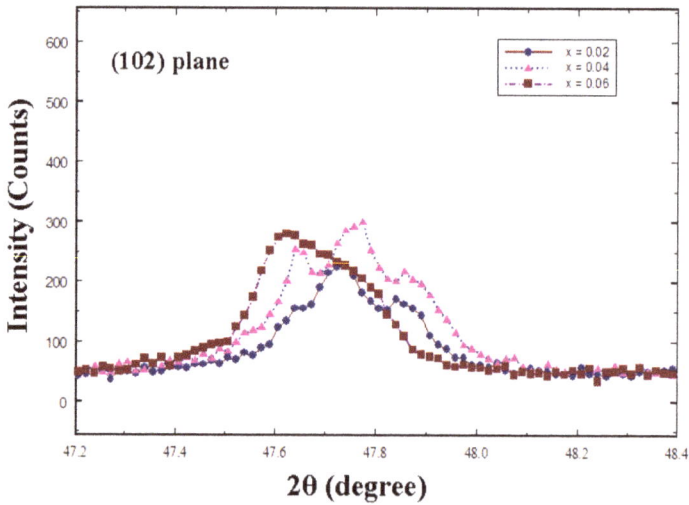

Figure 2.3 (e) Observed X-ray powder diffractograms of $Zn_{1-x}V_xO$ (x=0.02, 0.04, 0.06) at $2\theta=47.2°-48.4°$

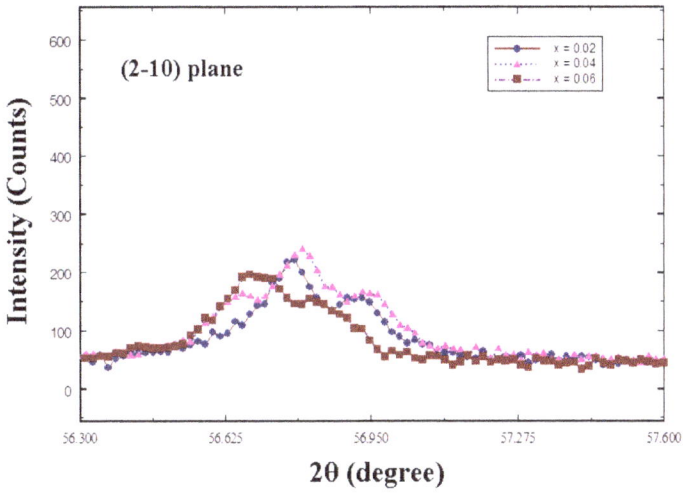

Figure 2.3 (f) Observed X-ray powder diffractograms of $Zn_{1-x}V_xO$ (x=0.02, 0.04, 0.06) at $2\theta=56.3^\circ$-57.6°

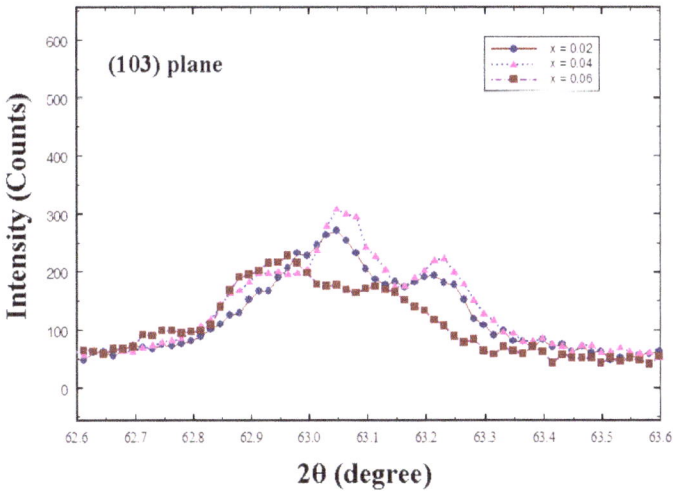

Figure 2.3 (g) Observed X-ray powder diffractograms of $Zn_{1-x}V_xO$ (x=0.02, 0.04, 0.06) at $2\theta=62.6^\circ$-62.6°

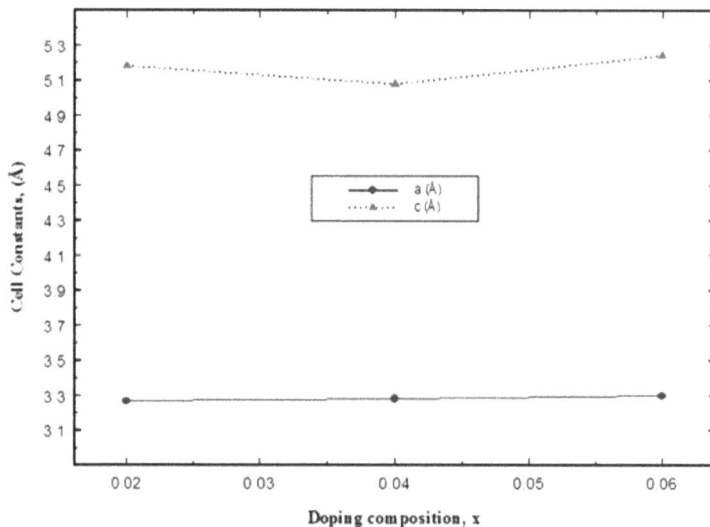

Figure 2.4 Vegard's plot for three different compositions of $Zn_{1-x}V_xO$

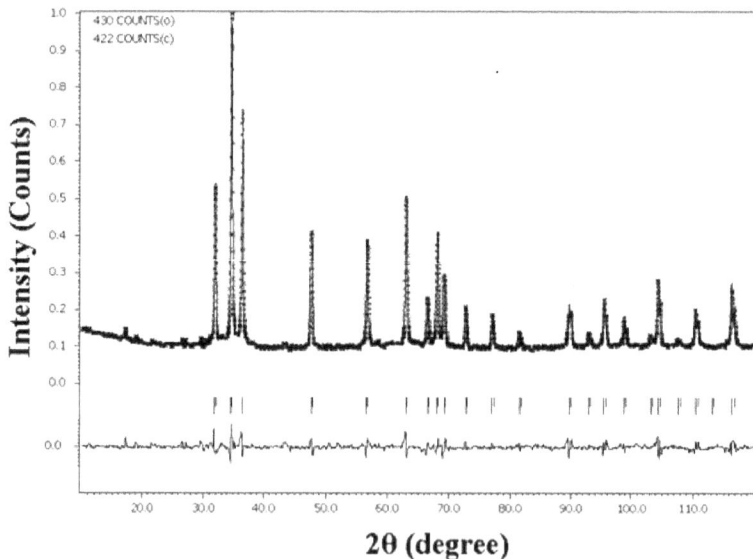

Figure 2.5 (a) Rietveld refined powder XRD profile for $Zn_{0.98}V_{0.02}O$

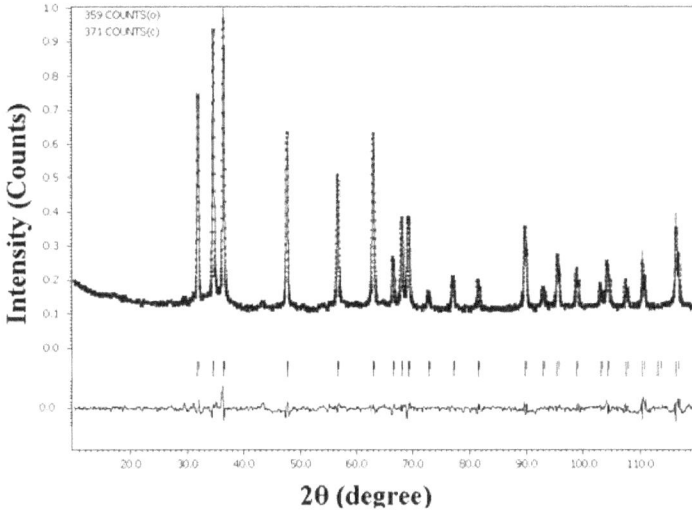

Figure 2.5 (b) Rietveld refined powder XRD profile for $Zn_{0.96}V_{0.04}O$

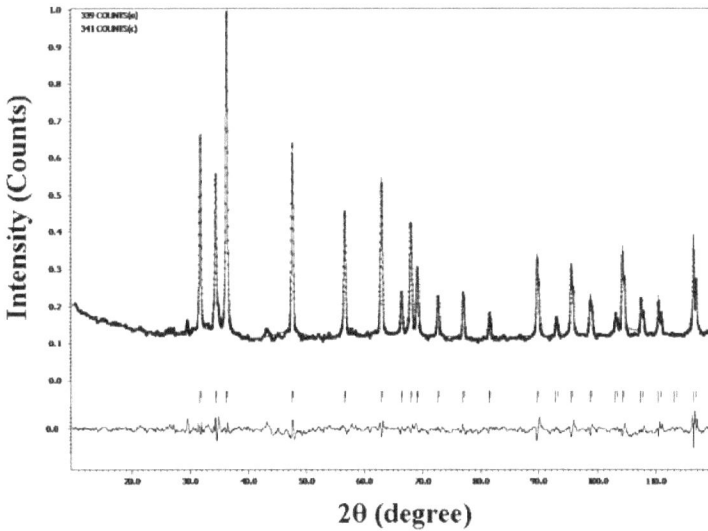

Figure 2.5 (c) Rietveld refined powder XRD profile for $Zn_{0.94}V_{0.06}O$

Table 2.8 XRD Bragg peak shifting of $Zn_{1-x}V_xO$

(hkl) plane	x=0.02	x=0.04	x=0.06	ZnO (*JCPDS)
(100)	31.95	32.01	31.88	31.77
(002)	34.68	34.65	34.48	34.42
(101)	36.44	36.47	36.37	36.25
(102)	47.72	47.78	47.61	47.53
(2-10)	56.72	56.82	56.3	56.60
(103)	62.04	62.06	62.95	62.86

*JCPDS reference number: 361451

Table 2.9 Structural parameters for $Zn_{1-x}V_xO$ from unit cell refinement

Parameter	x=0.02	x=0.04	x=0.06
a (Å)	2.2352	2.2377	2.2432
c (Å)	5.1846	5.1884	5.1972

Table 2.10 Refined reliability indices for $Zn_{1-x}V_xO$ from Rietveld refinement

		x=0.02	x=0.04	x=0.06
	R_{obs}(%)	7.85	8.65	10.73
	wR_{obs}(%)	6.77	11.05	10.86
$Zn_{1-x}V_xO$	R_p(%)	4.17	4.86	5.17
	wR_p(%)	6.06	7.90	8.12
	GOF	0.45	0.60	0.58

2.4.5 $Zn_{1-x}Ni_{x/2}V_{x/2}O$ (x=0.02, 0.04, 0.06)

The X-ray intensity data was collected with a 2θ ranging from 10° to 120° with 0.05 step size using a wavelength of 1.54056 Å. The Bragg peak shifts and peak intensities observed for ZnO planes (100), (002) and (101) are presented in Table 2.11. More observation explains the fact that the vanadium dopant was well substituted to the host ZnO lattice than that of NiO. The space group of V_2O_5 dopant was P*mmn* which was easily substituted in the space group of host Zinc oxide P63mc. Whereas NiO which is cubic (Fm-3m, space group) finds it difficult to get associated with the zinc oxide hexagonal wurtzite structure. Moreover, there is a shift in Bragg peaks over the left direction which was clearly depicted in figures 2.6 (a) and (b). Figures 2.6 (a) and (b) shows the presence of secondary phase of NiO peak at (111) and (002) *hkl* plane (JCPDS reference no: 471049). The secondary phase of NiO was not observed in x=0.02 concentration because of the low dopant concentration. As the dopant concentration increases the intensity of NiO peak also increases. Shift in peak positions towards lower 2θ values clearly indicates the incorporation V^{2+}, Ni^{2+} ions into the zinc host lattice. Since the atomic number of vanadium and nickel is lower than zinc, it facilitates the co-doping of lower atomic transition metal ions into that of Zn host element. This is consistent with the fact that the radius of Zn^{2+} (0.74 Å (Shannon, 1976)) is slightly larger than that of Ni^{2+} (0.69 Å (Shannon, 1976)) and V^{2+} (0.59 Å (Shannon, 1976)) ions. The average radius due to Ni and V ions will be (0.69 Å+0.59 Å) 0.64 Å. Therefore effective radius of guest ions is $(Ni/V)^{2+}$ ions is 0.64 Å which will be substituted in the Zn^{2+} ion having a radius of 0.74 Å which was reflected in Bragg peak shift. For the comparative function table 2.12 shows the obtained cell values through the unit cell refinement technique

The fitted profiles of the observed and calculated relative intensities along with their Bragg peaks of varying concentrations were depicted in figures 2.7 (a) - (c). The cell parameters and the intensities were refined using the software program JANA2006 (Petříček et al., 2006) by employing Rietveld techniques (Rietveld, 1969) taking into account the wurtzite structure of ZnO with space group of $P6_3mc$ (JCPDS reference no: 361451). The Rietveld refinement is a well-known powder profile fitting method for structural refinement. In this method, the observed profiles were matched with the similar profiles constructed by using pseudo-voigt (Wertheim et al., 1974) profile shape function of Thompson, Cox & Hastings (Thompson et al., 1987). The profile asymmetry parameter devised by Howard (Howard, 1982) was introduced by employing the multi term Simpson rule integration which incorporates symmetric profile shape function with different coefficient for weights and peak shift. The correction for preferred orientation was carried out using March-Dollase function (March et al., 1932, Dollase, 1986). The refined cell parameters and the final reliability indices were shown in Table 2.12 along with standard ZnO structural parameters (JCPDS reference no: 361451).

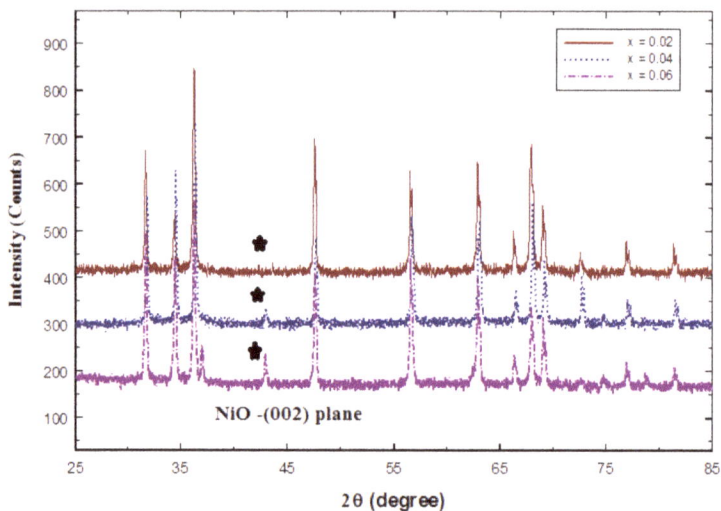

Figure 2.6 (a) Emerging NiO Bragg Peak in three compositions

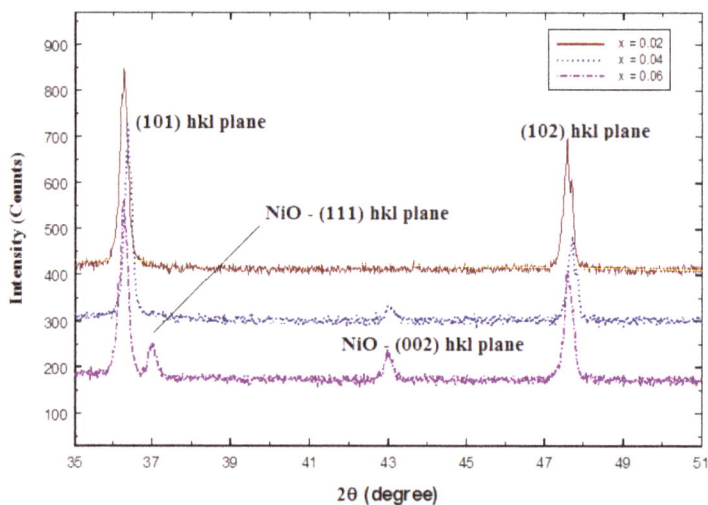

Figure 2.6 (b) Emerging NiO Bragg Peak in three compositions (e) Identified NiO Bragg peaks (111) hkl (002) hkl planes

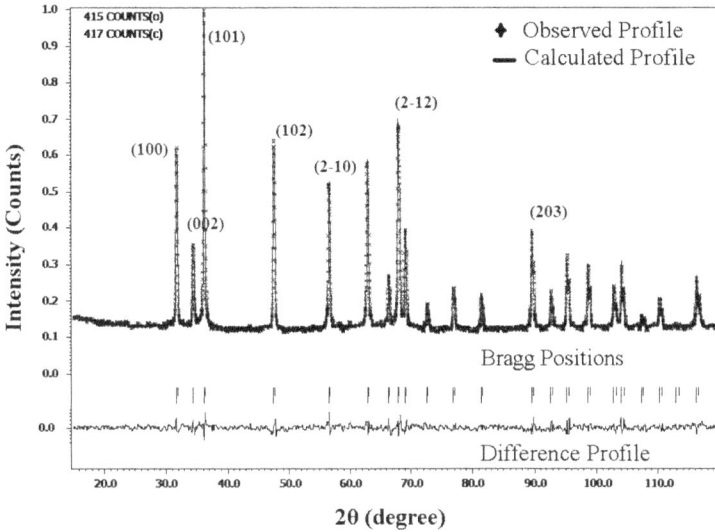

Figure 2.7 (a) Rietveld refined powder XRD profile for $Zn_{0.98}Ni_{0.01}V_{0.01}O$

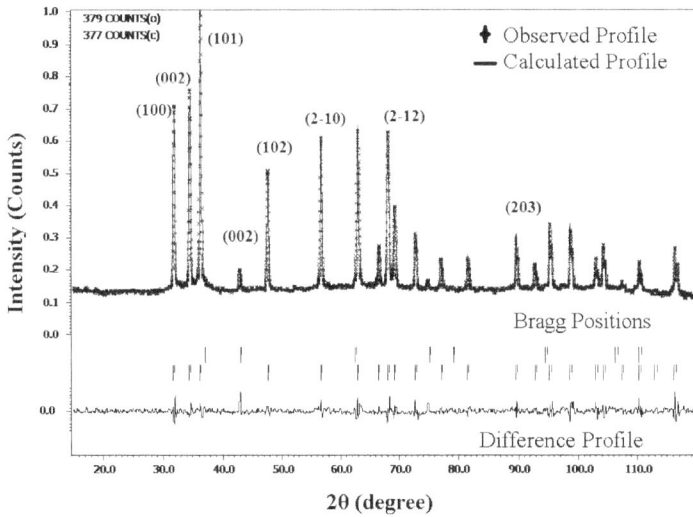

Figure 2.7 (b) Rietveld refined powder XRD profile for $Zn_{0.96}Ni_{0.02}V_{0.02}O$

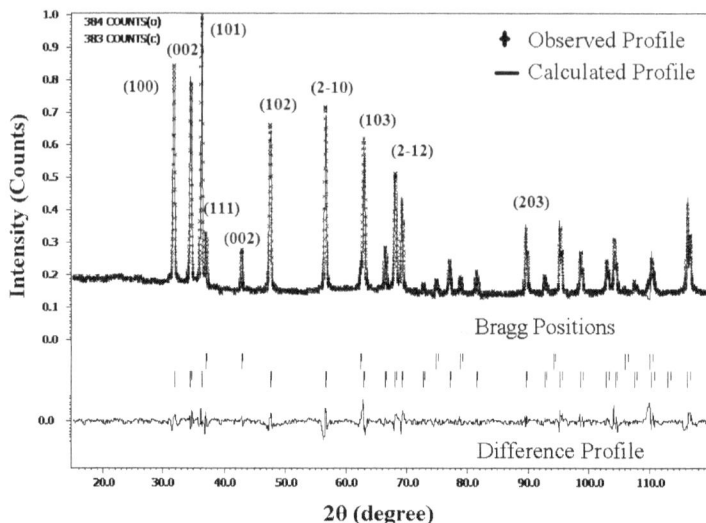

Figure 2.7 (c) Rietveld refined powder XRD profile for $Zn_{0.94}Ni_{0.03}V_{0.03}O$

Table 2.11 Bragg peak values for major (hkl) plane

	(hkl) plane	ZnO* (JCPDS)	x=0.02	x=0.04	x=0.06
$Zn_{1-x}Ni_{x/2}V_{x/2}O$	(100)	31.77	31.74	31.79	31.88
	(002)	34.42	34.42	34.48	34.67
	(101)	36.25	36.23	36.29	36.41

Table 2.12 Cell values through the unit cell refinement technique

		X= 0.02	X = 0.04	X = 0.06
$Zn_{1-x}Ni_{x/2}V_{x/2}O$	a(Å)	2.2520	2.2469	2.2503
	c (Å)	5.2055	5.1976	5.2002

Table 2.13 Structural parameters for $Zn_{1-x}Ni_{x/2}V_{x/2}O$ from Rietveld refinement

Parameter	x=0.02	x=0.04	x=0.06	ZnO*
a (Å)	2.2520(1)	2.2469(0)	2.2503(2)	2.249
c (Å)	5.2055(2)	5.1976(1)	5.2002(1)	5.206
Cell volume (Å^3)	47.67(1)	47.45(2)	47.57(1)	47.62
R_{obs}(%)	7.28	5.10	7.66	-
wR_{obs}(%)	7.67	5.08	8.70	-
R_p(%)	2.82	2.65	2.94	-
wR_p(%)	5.67	5.20	5.93	-

*JCPDS reference number PDF# 361451

The table 2.14. displays the relative values of all the raw XRD Bragg peaks for the major (hkl) planes of (100), (002) and (101) only and samples grown with various concentrations shows the dopant has been successfully implanted on the host lattice.

Table 2.14 A comparative table of raw XRD Bragg peaks of the grown samples

Samples	(hkl) plane	ZnO* (JCPDS)	x=0.01	x=0.02	x=0.03
$Zn_{1-x}Ti_xO$	(100)	31.77	31.77	31.75	31.67
	(002)	34.42	34.42	34.39	34.27
	(101)	36.25	36.25	36.19	36.15
			x=0.02	x=0.04	x=0.06
$Zn_{1-x}Fe_xO$	(100)	31.77	31.51	31.68	32.18
	(002)	34.42	32.97	34.23	34.62
	(101)	36.25	35.95	36.23	36.58
			x=0.02	x=0.04	x=0.06
$Zn_{1-x}V_xO$	(100)	31.77	31.95	32.01	31.88
	(002)	34.42	34.68	34.65	34.48
	(101)	36.25	36.44	36.47	36.37
			x=0.02	x=0.04	x=0.06
$Zn_{1-x}Ni_{x/2}V_{x/2}O$	(100)	31.77	31.74	31.79	31.88
	(002)	34.42	34.42	34.48	34.67
	(101)	36.25	36.23	36.29	36.41

An idea regarding how the dopant has affected the cell parameters of the host atom the table 2.15 has been drawn showing refined cell parameters alone for all sample with its own varying concentrations.

Table 2.15 A comparative table of refined cell parameters of the grown samples

Composition	ZnO*	X = 0.01	x=0.02	x=0.03
	a(Å) = 2.249	2.2470	2.2478	2.2500
$Zn_{1-x}Ti_xO$	c (Å) = 5.206	5.2034	5.2030	5.2072
	Cell volume (Å^3) = 47.62	47.5	47.5	47.6
		X =0.02	X = 0.04	X = 0.06
	a(Å) = 2.249	2.2493	2.2539	2.2539
$Zn_{1-x}Fe_xO$	c (Å) = 5.206	5.2092	5.2313	5.2208
	Cell volume (Å^3) = 47.62	47.6	48.0	47.9
		X= 0.02	X = 0.04	X = 0.06
	a(Å) = 2.249	2.2527	2.2505	2.2490
$Zn_{1-x}V_xO$	c (Å) = 5.206	5.2059	5.2507	5.2066
	Cell volume (Å^3) = 47.62	47.7	47.6	47.6
		X = 0.02	X = 0.04	X = 0.06
	a(Å) = 2.249	2.2520	2.2469	2.2503
$Zn_{1-x}Ni_{x/2}V_{x/2}O$	c (Å) = 5.206	5.2055	5.1976	5.2002
	Cell volume (Å^3) = 47.62	47.67	47.45	47.57

The refined cell parameters and the final reliability indices were shown in Table 2.16.

Table 2.16 Refined parameters through Rietveld procedures

Composition	Refinement parameters	x=0.01	x=0.02	x=0.03
	R_{obs}(%)	8.67	6.53	5.51
$Zn_{1-x}Ti_xO$	wR_{obs}(%)	8.59	8.09	7.09
	R_p(%)	11.55	8.76	8.94
	wR_p(%)	15.09	11.38	11.77

	GOF	1.23	0.92	0.97
		x=0.02	x=0.04	x=0.06
$Zn_{1-x}Fe_xO$	$R_{obs}(\%)$	2.55	0.91	1.56
	$wR_{obs}(\%)$	2.02	1.11	1.76
	$R_p(\%)$	6.18	6.63	6.42
	$wR_p(\%)$	8.13	9.05	8.29
	GOF	1.64	1.71	1.64
		x=0.02	x=0.04	x=0.06
$Zn_{1-x}V_xO$	$R_{obs}(\%)$	7.85	8.65	10.73
	$wR_{obs}(\%)$	6.77	11.05	10.86
	$R_p(\%)$	4.17	4.86	5.17
	$wR_p(\%)$	6.06	7.90	8.12
	GOF	0.45	0.60	0.58
		x=0.02	x=0.04	x=0.06
$Zn_{1-x}Ni_{x/2}V_{x/2}O$	$R_{obs}(\%)$	7.28	5.10	7.66
	$wR_{obs}(\%)$	7.67	5.08	8.70
	$R_p(\%)$	2.82	2.65	2.94
	$wR_p(\%)$	5.67	5.20	5.93
	GOF	0.45	0.41	0.56

R_{obs} – Reliability index for observed profile
R_p - Reliability index for profile

References

[1] Braga, D., Desiraju, G.R., Miller, J.S., Orpen, A.G., Price, S.L., (2002), Innovation in crystal engineering, Crystal engineering communications, 4, 500-509. https://doi.org/10.1039/B207466B

[2] Callister, Jr., William, D., (2007), Materials Science and Engineering an Introduction, 7th Edition, John Wiley and Sons, New York.

[3] Chikoidze E., Dumont Y., Jomard F., Ballutaud D., Galtier P., Gorochov O., Ferrand D., J. Appl. Phys. 97, D327 (2005). https://doi.org/10.1063/1.1863132

[4] D. Bradley, (2002), Nano the Hedgehog, The Alchemist, News Research.

[5] Desiraju, G.R., (1989), Crystal Engineering: The Design of Organic Solids, Elsevier, New York.

[6] Dollase W.A.J., Appl. Cryst. 19, 267 (1986). https://doi.org/10.1107/S0021889886089458

[7] Holland T.J.B., Redfern S.A.T., Mineral. Mag. 61, 65-77 (1997). https://doi.org/10.1180/minmag.1997.061.404.07

[8] Howard C.J., J. Appl. Cryst. 15, 615 (1982).

https://doi.org/10.1107/S0021889882012783

[9] Karamat S., Rawat R.S., Lee P., Tan T.L., Ramanujan R.V., Zhou W., Applied Surface Science, 256, 2309-2314 (2010). https://doi.org/10.1016/j.apsusc.2009.09.039

[10] March, Z. Kristallogr. 81, 285 (1932). https://doi.org/10.1524/zkri.1932.81.1.285

[11] Narayan H., Alemu H., Macheli L., Rao G., Nanotechnology. 20, 255601 (2009). https://doi.org/10.1088/0957-4484/20/25/255601

[12] Petříček V., Dušek M., Palatinus L., JANA 2006, The crystallographic computing system Institute of Physics Academy of sciences of the Czech republic, Praha (2006).

[13] Rietveld H.M., J. Appl. Crystallogr., 2, 65 (1969). https://doi.org/10.1107/S0021889869006558

[14] Shannon R.D., Acta Cryst. A32, 751-767 (1976). https://doi.org/10.1107/S0567739476001551

[15] Singh S., Rama N., Sethupathi K., Ramachandra Rao M.S., J. Appl. Phys. 103(7), 07D108-1-07D108-3) (2008). https://doi.org/10.1063/1.2834443

[16] Stanley F. Bartam, Richard A. Slepetys, J. Am. Ceram. Soc. 44(10), 493-499 (2011). https://doi.org/10.1111/j.1151-2916.1961.tb13712.x

[17] Thompson P., Cox D.E., Hastings J.B., J. Appl. Cryst. 20, 79 (1987). https://doi.org/10.1107/S0021889887087090

[18] Wertheim G.K., Butler M.A., West K.W., Buchanan D.N.E., Rev. Sci. Instrum. 45, 1369 (1974). https://doi.org/10.1063/1.1686503

[19] Wu P., Saraf G., Lu Y., Hill D.H., Gateau R., Wielunski L., Bartynsk R.A., Arena D.A., Dvorak J., Moodenbaugh A., Siegrist T., Raley J.A., Yeo Y.K., Appl. Phys. Lett. 89, 012508 (2006). https://doi.org/10.1063/1.2213519

Chapter 3

Surface Morphological, Optical, Magnetic Properties of DMS Materials

Abstract

The oxide based dilute magnetic semiconductors synthesized in the present work were characterized for

- Structural and average particle size analysis using X-ray diffraction studies.

- Morphological analysis using scanning electron microscope (SEM) studies.

- Elemental analysis using energy dispersive X-ray (EDX) studies.

- Magnetic properties analysis using vibrating sample magnetometer (VSM) studies.

- Optical properties using UV-Vis characterization studies.

Room temperature raw X-ray diffraction of all powder ferrite samples were recorded at NIIST (National Institute for Interdisciplinary Science and Technology, Council of Scientific and Industrial Research (CSIR)), Thiruvananthapuram, India, using an X-PERT PRO (Philips, Netherlands) X-ray diffractometer. The monochromator used for X-ray diffraction was copper which produces $CuK\alpha1$ radiation as incident beam. The wavelength used for the X-ray intensity data collection was 1.54056 Å with a 2θ range of data collection from 10° to 120° with 0.05° step size. The morphological analysis of the powder samples was studied with scanning electron microscope (SEM) (Hitachi, Model: S-3400 N) at Central Instrumentation Facility (CIF), Pondicherry University, India.

Energy Dispersive X-Ray (EDX) analyses of all pelletized samples, were recorded at Sophisticated Analytical Instruments Facility (SAIF), IIT-Madras, India, using Model: Quanta 200 FEG. Room temperature magnetic properties of all powder samples were recorded at SAIF, IIT-Madras, India, using VSM (Model: LakeshoreVSM 7410). The UV-Vis characterization of all samples under investigation in this research work, in order to evaluate the optical band gap energy, was carried out by using UV-Vis-NIR Spectrophotometer (Model: Varian, Cary 5000) at STIC, CUSAT, Cochin, India. The range of wavelength employed was 200 – 800nm.

Keywords

Optical, Morphological, Magnetic, Compositional, Structural, SEM, EDX, VSM, X-Ray

3.1 Introduction

The surface morphology of sintered samples has been analyzed through scanning electron microscopy (SEM). The elemental compositions of the solid solution samples have been analyzed by energy dispersive X-ray spectroscopy (EDS). The optical properties of the samples have been investigated by UV-visible spectroscopy. Band gap values of prepared samples have been evaluated extrapolating the linear portion of the Tauc plot. Magnetic properties of the grown samples have been measured and analyzed.

3.2 Surface morphological properties of DMS materials

3.2.1 $Zn_{1-x}Ti_xO$ (x=0.01, 0.02, 0.03)

The surface morphology of the powders was investigated with the help of scanning electron microscope. Figures 3.1 (a) and (b) show the SEM micrographs of the grown $Zn_{1-x}Ti_xO$ (x = 0.01, x\= 0.02, x = 0.03) powder samples showing randomly distributed crystallites. Crystallite size measured using the Scherrer's equation (Samavati et al., 2016) was found to increase with the concentration of dopant from 23 to 26 nm as shown in Table 3.1. As the doping concentration increases, the size of the particles increases, which are in line with increase in lattice parameter values (Table 2.3). The incorporation of Ti ions at the Zn lattice site seems to influence the surface morphology of the samples. The average particle size of the grown sample determined from SEM micrograph also show increasing trend with the concentration x as shown Table 3.1.

(a)   **(b)**

(c)

Figure 3.1 SEM micrographs of $Zn_{1-x}Ti_xO$ samples (a) x=0.01 (b) x=0.02 (c) x=0.03

Table 3.1 Crystallite size and particle size for $Zn_{1-x}Ti_xO$ (x=0.02, 0.03) samples

Concentrations	Crystallite size (nm)	Particle size (µm)
x = 0.01	23.439	2.010
x = 0.02	24.412	2.025
x = 0.03	26.965	2.675

3.2.2 $Zn_{1-x}Fe_xO$ (x=0.02, 0.04, 0.06)

The micro structure measurements were carried out under different magnifications using scanning electron microscopy for the system $Zn_{1-x}Fe_xO$ (x=0.02, 0.04, 0.06) are shown in figures 3.2 (a)-(c). Different grain sizes are heterogeneously distributed throughout the samples. The average grain size for the three compositions determined to be 1.58 µm, 2.31 µm and 2.03 µm for x=0.02, 0.04, 0.06 respectively. In all the compositions, the grain sizes distributed heterogeneously throughout the samples. A comparison were also done with crystallite size of the samples using GRAIN software (Saravanan, 2008) and presented in Table 3.2. Increase in crystallite size as the dopant concentration.

The inclusion of ferrous dopant on all three compositions of the system was confirmed further by the elemental analysis carried out using energy dispersive X-ray spectroscopy (EDS). The elemental analysis by energy dispersive X-ray spectroscopy (EDS) confirms the presence of ferrous dopant, in all the three compositions, as shown in figures 3.3 (a)-(c).

Figure 3.2 FESEM micrographs of $Zn_{1-x}Fe_xO$ samples (a) x=0.02 (b) x=0.04 (c) x=0.06

Table 3.2 Crystallite size and particle size for $Zn_{1-x}Fe_xO$ (x=0.02, 0.04, 0.06) samples

Concentrations	Crystallite size (nm)	Particle size (µm)
x=0.02	26.47	1.90
x=0.04	76.19	2.25
x=0.06	85.02	2.77

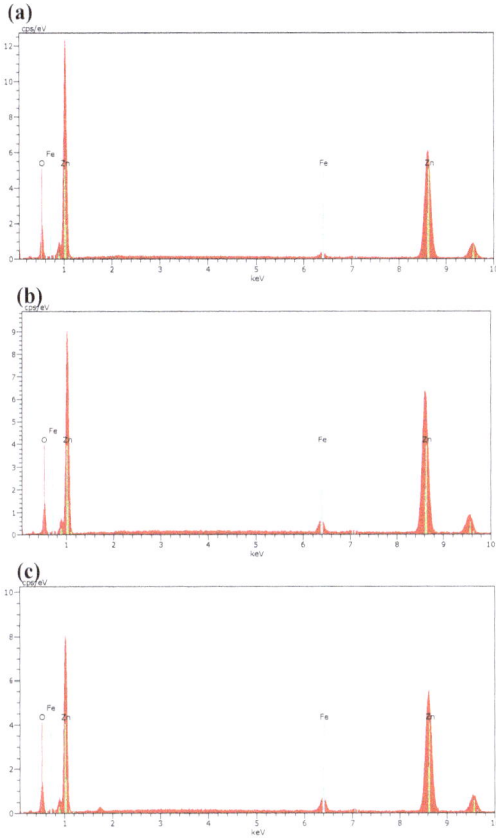

Figure 3.3 EDS spectra of $Zn_{1-x}Fe_xO$ samples (a) x=0.02 (b) x=0.04 (c) x=0.06

3.2.3 $Zn_{1-x}V_xO$ (x=0.02, 0.04, 0.06)

The microstructure measurements were carried out using field emission scanning electron microscopy for the system $Zn_{1-x}V_xO$ which are shown in figures 3.4 (a)-(c). Different grain sizes are heterogeneously distributed throughout the samples. The average grain size for the three compositions determined to be 1.58 μm, 2.31 μm and 2.03 μm for x=0.02, 0.04 and x=0.06 respectively. A comparative table along with crystallite size is displayed in table 3.3. The elemental analysis by energy dispersive X-ray spectroscopy (EDS) confirms the presence of vanadium dopant, in all three compositions, as shown in figures 3.5 (a)-(c).

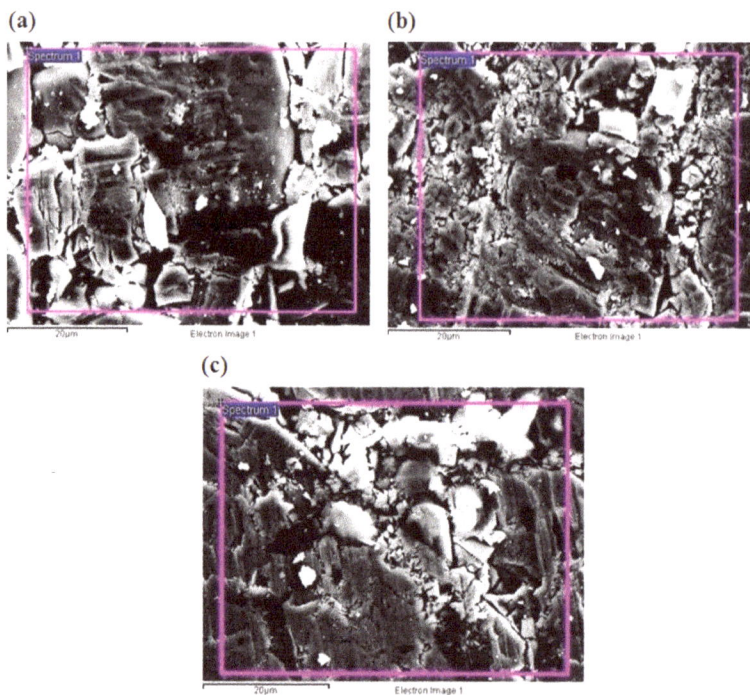

Figure 3.4 FESEM micrographs of $Zn_{1-x}V_xO$ samples (a) x=0.02 (b) x=0.04 (c) x=0.06 concentrations

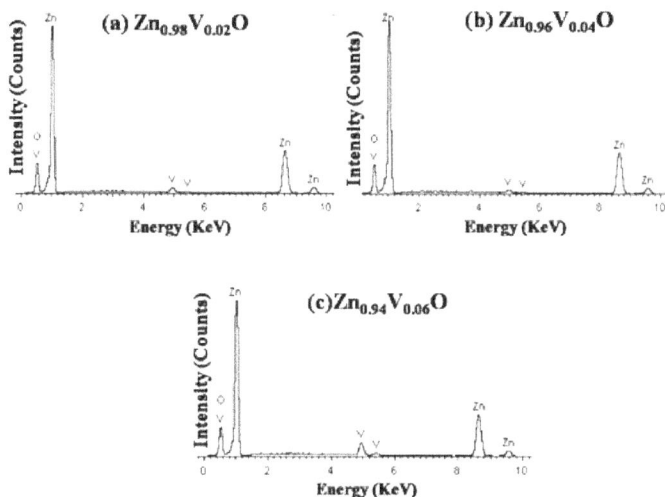

Figure 3.5 EDS spectra of $Zn_{1-x}V_xO$ samples (a) x=0.02 (b) x=0.04 (c) x=0.06 concentrations

Table 3.3 Crystallite size for three compositions of $Zn_{1-x}V_xO$

Concentrations	Crystallite size (nm)	Particle size (μm)
x=0.02	22.9485	1.58
x=0.04	23.8513	2.31
x=0.06	27.5471	2.03

3.2.4 $Zn_{1-x}Ni_{x/2}V_{x/2}O$ (x=0.02, 0.04, 0.06)

The particle size of the grown system with varying dopant concentration was evaluated using GRAIN (Saravanan, 2008) software. The size of the Ni and V co-doped system with varying concentrations is analyzed using full width at half maximum of the powder XRD peaks. From this analysis the crystallite size (r_{Xray}) comes out to be in the range of 22.94 nm to 27.54 nm presented in Table 3.4. Increasing trend was observed in particle size as the dopant concentration increases as shown in table 3.4, which was later similar with ICP-AES studies. Inductively coupled plasma emission spectroscopy was employed to analyze the elements with respect to their composition and their values are tabulated in Table 3.5.

Table 3.4 Crystallite size for three compositions of $Zn_{1-x}Ni_{x/2}V_{x/2}O$

Concentrations	Crystallite size (nm)	Particle size (µm)
x=0.02	22.9485	1.31
x=0.04	23.8513	1.95
x=0.06	27.5471	2.32

Table 3.5 Elemental detection through ICP-AES analysis of $Zn_{1-x}Ni_{x/2}V_{x/2}O$

Concentrations	Zinc	Vanadium	Nickel	Unit %	Dopant Concentration
x=0.02	88.77	0.05	1.91	%	1.96
x=0.04	80.59	0.02	3.80	%	3.82
x=0.06	79.69	0.05	1.03	%	1.08

Complete morphological studies of all grown samples were completed and its crystallite size and average particle size were studies as discussed earlier. Table 3.6 gives the correlated values of the grown samples morphological data.

Table 3.6 Comparative tables for crystallite size of grown samples

Sample	Composition	Crystallite size (nm)	Particle Size (µm)
	x = 0.01	23.4391	2.01
$Zn_{1-x}Ti_xO$	X= 0.02	23.4123	2.02
	X= 0.03	26.9651	2.67
$Zn_{1-x}Fe_xO$	X= 0.02	26.4713	1.90
	X= 0.04	76.1945	2.25
	X= 0.06	85.0258	2.77
$Zn_{1-x}V_xO$	X= 0.02	22.9485	1.58
	X= 0.04	23.8513	2.31
	X= 0.06	27.5471	2.03
$Zn_{1-x}Ni_{x/2}V_{x/2}O$	X= 0.02	22.9485	1.31
	X= 0.04	23.8513	1.95
	X= 0.06	27.5471	2.32

3.3 Optical properties of DMS materials

3.3.1 Band gap from optical absorption spectra of $Zn_{1-x}Ti_xO$ (x=0.02, 0.03)

UV-vis absorption spectra were recorded and the optical band gap was obtained by extrapolating of the linear part until it intersects with x-axis. Figure 3.6 (a) shows the dependence of $(\alpha hv)^{1/2}$ on photon energy hv. The optical band gap energy of the $Zn_{1-x}Ti_xO$ system was depicted in Table 3.7. It is confirmed that the band gaps of Ti-doped ZnO increase with increase in doping concentration of titanium as in V-doped ZnO films (Luo et al., 2009). The increase in band gap energy may be attributed to Burstein-Moss effect. The addition of Ti merges the Fermi level into the conduction band.

3.3.2 Photoluminescence studies of $Zn_{1-x}Ti_xO$ (x=0.02, 0.03)

Photoluminescence (PL) measurements were observed in pelletized samples of Ti doped ZnO system using ISS - Absorption/Fluorescence spectrophotometer fitted with a high power excitation xenon lamp. The observed Photoluminescence spectra were presented in figure 3.6 (b). The excitation region was in the region of 320 nm region and emission occurs in the region of 500-650 nm for both the concentrations. The emission spectrum for the intrinsic luminescence band of ZnO was at 545 nm. The presence of dopant showed an observable red-shift when compared with emission of intrinsic band of undoped ZnO (Kuznetsov et al., 2012). The emission spectra shown in the figure demonstrates the contributions of dopant concentrations to the intensity of the observed emission bands.

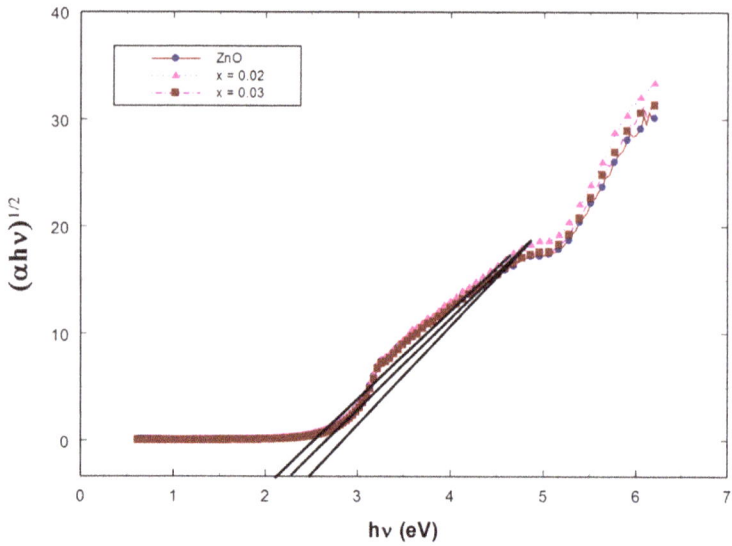

Figure 3.6 (a) Tauc plot for $Zn_{1-x}Ti_xO$ (x=0.00, 0.02, 0.03) samples

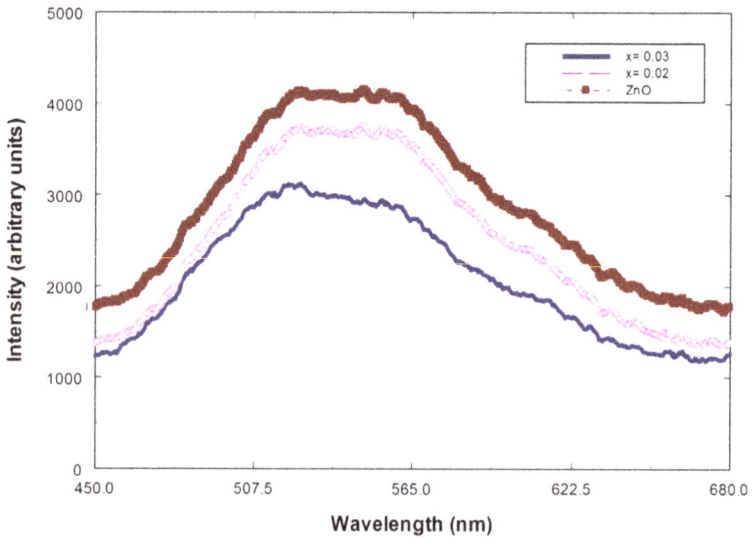

Figure 3.6 (b) PL Emission spectra of $Zn_{1-x}Ti_xO$ (x=0.00, 0.02, 0.03) samples

Table 3.7 Direct band gap values for $Zn_{1-x}Ti_xO$ (x=0.01, 0.02, 0.03) samples

Concentrations	Band gap (eV)
x=0.01	2.68
x=0.02	2.45
x=0.03	2.89

3.3.3 Optical studies on $Zn_{1-x}Fe_xO$ (x=0.02, 0.04, 0.06)

The grown $Zn_{1-x}Fe_xO$ (x=0.02, 0.04, 0.06) samples were analyzed for direct band gap values and Tauc Plot was plotted (Wood and Tauc, 1972). The adsorption spectra shows an increasing trend with respect to increasing concentration levels, which was shown in figure 3.7 and the results ranges from 3.21 eV, 3.43 eV and 2.86 eV for x = 0.02, x = 0.04, x = 0.06 concentrations respectively, table 3.8. Among these direct band gap values x = 0.04 concentration doping coincides well with reported zinc oxide value of host (3.37 eV). The direct band gap values changes were attributed to the changes in the coercivity and saturation magnetization values for the compositions grown shown in table 3.12.

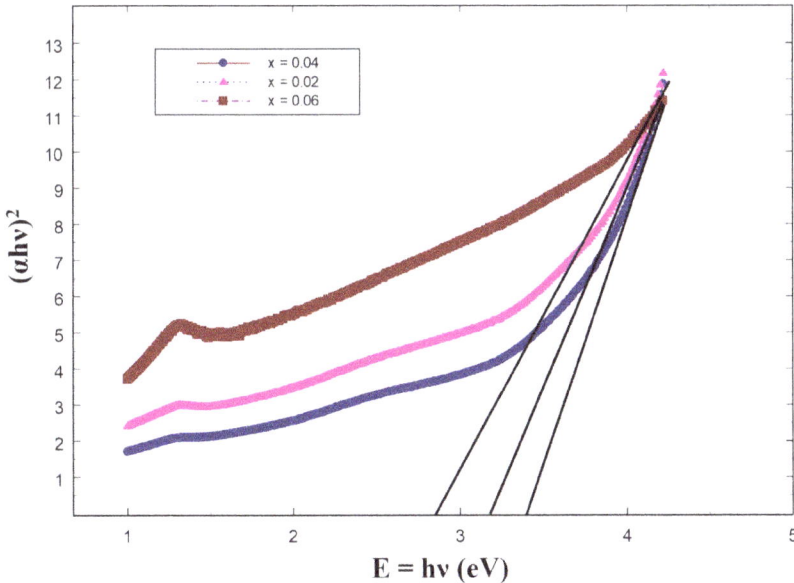

Figure 3.7 UV –Vis. Spectra for $Zn_{1-x}Fe_xO$ (x=0.02, 0.04, 0.06)

Table 3.8 Direct band gap values for $Zn_{1-x}Fe_xO$ (x=0.02, 0.04, 0.06) samples

Concentrations	Band gap (eV)
x=0.02	3.21
x=0.04	3.43
x=0.06	2.86

3.3.4 Optical band gap studies for $Zn_{1-x}V_xO$ (x = 0.02, x = 0.04, x = 0.06)

UV-Vis spectra in figure 3.8 for the systems $Zn_{1-x}V_xO$ (where x = 0.02, x = 0.04, x = 0.06) was plotted as the peak of absorbance spectra ($(\alpha h\nu)^2$ *vs.* $h\nu$ plot), where α is the absorption co-efficient and ν the photon frequency. The direct band gap for the system $Zn_{1-x}V_xO$ (where x = 0.02, x = 0.04, x = 0.06) were calculated using Tauc plot (Wood and Tauc, 1972). The optical band gap values obtained are in the range, table3.9, for x = 0.02, x= 0.04 and x = 0.06 compositions were 3.12 eV, 3.23 eV and 3.41 eV indicating that Vanadium is properly substituted in the matrix thereby showing an appreciable increase compared with the band gap value of ZnO of 3.37 eV.

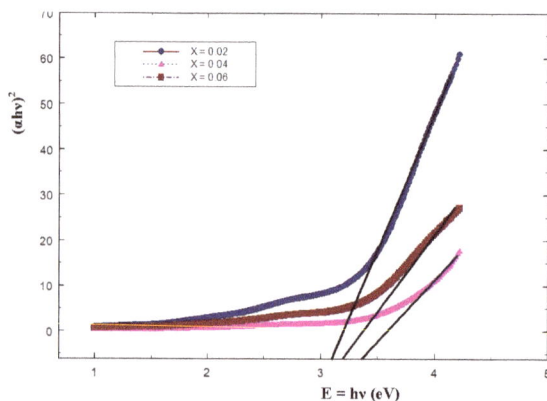

Figure 3.8 Optical spectra for $Zn_{1-x}V_xO$ (x=0.02, 0.04, 0.06)

Table 3.9 Direct band gap values for $Zn_{1-x}V_xO$ (x=0.02, 0.04, 0.06) samples

Concentrations	Band gap (eV)
x=0.02	3.12
x=0.04	3.23
x=0.06	2.41

3.3.5 UV-visible spectra-optical band gap studies of $Zn_{1-x}Ni_{x/2}V_{x/2}O$ (x=0.02, 0.04, 0.06)

Optical property of the grown samples was analysed through UV-vis spectra. The direct band gap of the material was calculated by drawing Tauc plot (Wood and Tauc, 1972) with respect to hv (eV) as x-axis and $(\alpha hv)^{1/2}$ as y-axis shown in figure 3.9. Their corresponding direct band gap values were shown in Table 3.10. The values in Table 3.10 shows an increasing trend as the dopant concentration decreases ranging from 1.89 eV to 1.62 eV. This may be attributed to the fact that any addition of metal decreases the value of direct band gap value of zinc oxide (3.37 eV).

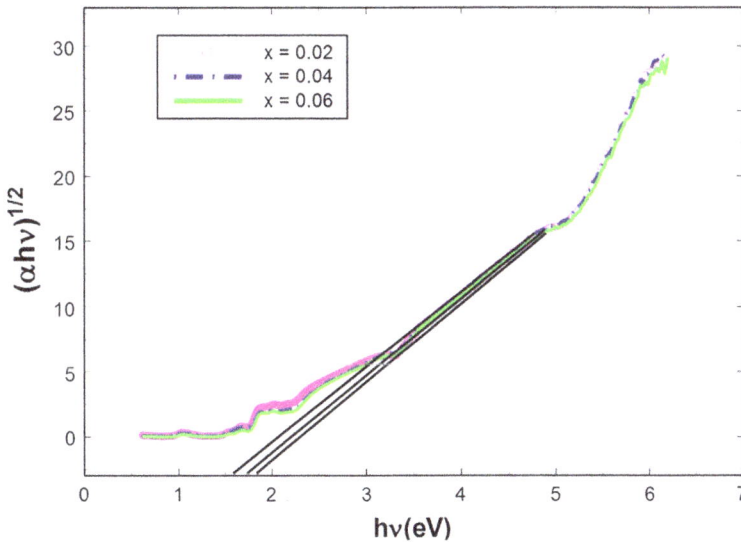

Figure 3.9 Tauc plot for $Zn_{1-x}Ni_{x\,2}V_{x\,2}O$ (x=0.02, 0.04, 0.06) samples

Table 3.10 Band values for $Zn_{1-x}Ni_{x\,2}V_{x\,2}O$ (x=0.02, 0.04, 0.06) samples

Concentrations	Band gap (eV)
x=0.02	1.89
x=0.04	1.70
x=0.06	1.62

An overall comparison for all the direct band gap values for the grown samples was depicted in table 3.11. It may give an idea related to the optical studies for the grown samples

Table 3.11 Relative band gap values for grown samples

Sample	Composition	Band Gap (eV)
$Zn_{1-x}Ti_xO$	X= 0.01	2.68
	X= 0.02	2.45
	X= 0.03	2.89
$Zn_{1-x}Fe_xO$	X= 0.02	3.21
	X= 0.04	3.43
	X= 0.06	2.96
$Zn_{1-x}V_xO$	X= 0.02	3.12
	X= 0.04	3.23
	X= 0.06	3.41
$Zn_{1-x}Ni_{x/2}V_{x/2}O$	X= 0.02	1.80
	X= 0.04	1.70
	X= 0.06	1.62

3.4 Magnetic properties of DMS materials

3.4.1 $Zn_{1-x}Ti_xO$ (x=0.01, x=0.02, 0.03)

The magnetic behaviour of transition metal Ti doped ZnO bulk samples has been studied at RT, because it is obvious that the growth conditions decide the magnetic properties in a system. Figure 3.10 (a) shows the magnetic field versus the magnetic moment curves with room temperature ferromagnetism. The close-up view of figure 3.10 (a) near H=0 Oe is shown in figure 3.10 (b). The saturation magnetization (M_s) (Table 3.12) increases with addition of Ti atoms. The difference in the M_s values between x=0.02 and 0.03 is 1.0715 (emu/g) indicating an even addition of Ti in these compositions. The observed magnetic behaviour in the samples is due to the inclusion of Ti^{4+} ions in the lattice sites of Zn^{2+} (Singh et al., 2008).

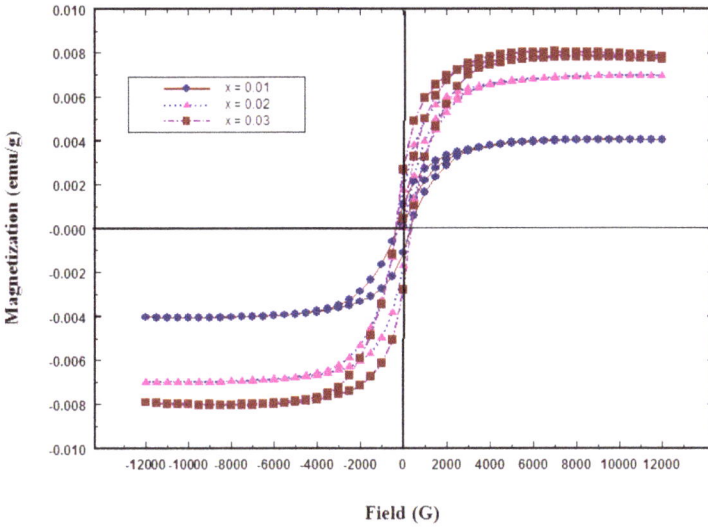

Figure 3.10 (a) M-H curves for $Zn_{1-x}Ti_xO$ (x = 0.01, 0.02, 0.03) samples

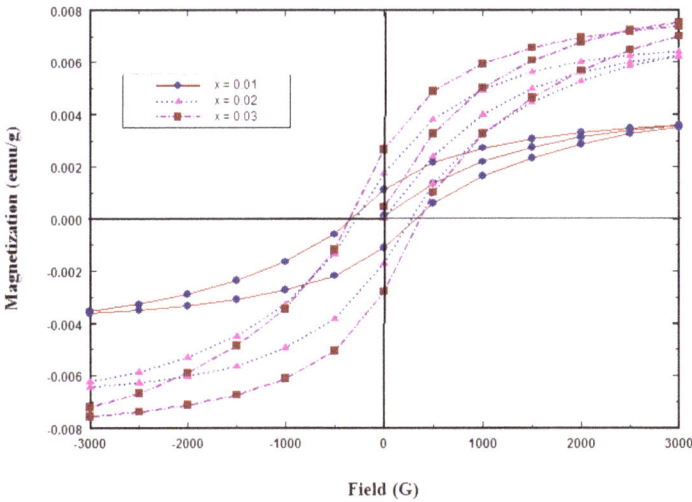

Figure 3.10 (b) A close view of M-H hysteresis curve for $Zn_{1-x}Ti_xO$ (x = 0.01, 0.02, 0.03) samples

Table 3.12 Magnetic measurements for $Zn_{1-x}Ti_xO$ (x = 0.01, 0.02, 0.03) samples

Concentrations	Coercivity (G)	Saturation Magnetization M_s (emu/g) ($x10^{-3}$)	Retentivity (emu/g) ($x10^{-3}$)
x=0.01	278.56	5.4856	0.8563
x=0.02	280.43	6.9929	1.7261
x=0.03	356.08	8.0644	2.7222

3.4.2   Magnetic measurements of $Zn_{1-x}Fe_xO$ (x=0.02, 0.04, 0.06)

The magnetic characterizations were done for all the samples by using vibrating sample magnetometery (VSM). Magnetic behavior of the materials $Zn_{1-x}Fe_xO$ for three different composition, x=0.02, 0.04 and 0.06 are presented in figures 3.11 (a) and (b). The hysteresis curves were traced for all the three concentrations confirm room temperature ferromagnetic behavior for all the samples as depicted in Table 3.13. The presence of Fe^{2+} in ZnO host lattice and the magnetic exchange interaction between magnetic moments in the sample contributes towards the ferromagnetic state (Krithiga et al., 2011). The observed order of saturation magnetization may be due to the magneto crystalline anisotropy (Krithiga et al., 2011) of the samples. The trend can also be observed in microstructure measurements.

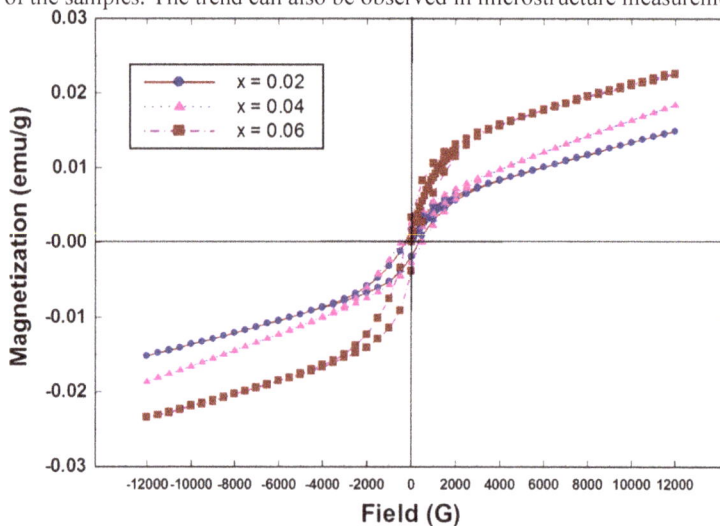

Figure 3.11 (a) M-H curves for $Zn_{1-x}Fe_xO$ (x=0.02, 0.04, 0.06) samples

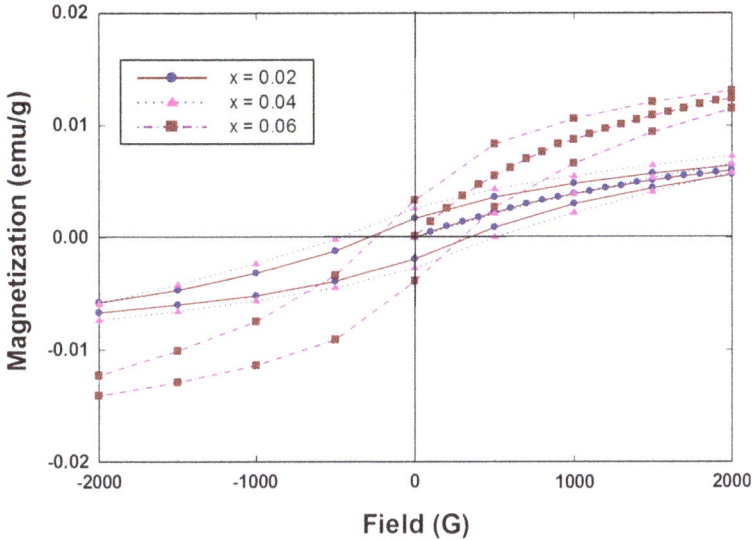

Figure 3.11 (b) Enhanced view of the hysteresis loop of $Zn_{1-x}Fe_xO$ (x=0.02, 0.04, 0.06) samples

Table 3.13 Magnetic parameters for $Zn_{1-x}Fe_xO$ (x=0.02, 0.04, 0.06) samples

Concentrations	Coercivity (G)	Saturation Magnetization Ms(emu/g) (x10^{-3})	Retentivity (emu/g) (x10^{-3})
x=0.02	317.97	15.083	1.816
x=0.04	481.46	18.537	2.666
x=0.06	273.04	23.031	3.631

3.4.3 Magnetic measurements of $Zn_{1-x}V_xO$ (x=0.02, 0.04, 0.06)

The magnetic characterizations were done for all the samples by using vibrating sample magnetometery (VSM). Magnetic behavior of the materials $Zn_{1-x}V_xO$, for three different composition i.e., x=0.02, 0.04 and 0.06 are shown in figure 3.12 (a) and enhanced view in figure 3.12(b). The hysteresis curves traced for all the three concentrations confirm room temperature ferromagnetic behavior for all the samples and it is more evident in x=0.04 concentration as depicted in Table 3.14. The high magnetic saturation in x=0.04 concentration may be attributed to the covalent nature as evidenced from the values of mid-bond electron densities presented in Table 4.3 and one dimensional charge density profiles are shown in figures 4.7 (a)-(c). The one dimensional electron density values (Table 4.3) for x=0.04 concentration show bond 1 is highly covalent and bond 2 is ionic. High covalent values and low ionic values in the composition enable the ferromagnetic property at x=0.04 to be highest. The presence of V^{2+} in ZnO host lattice and the magnetic exchange interaction between magnetic moments in the sample contributes towards the ferromagnetic state (Krithiga et al., 2011). The observed order of saturation magnetization may be due to the magneto crystalline anisotropy of the samples. The trend where x=0.04 concentration is greater can also be observed in microstructure measurements.

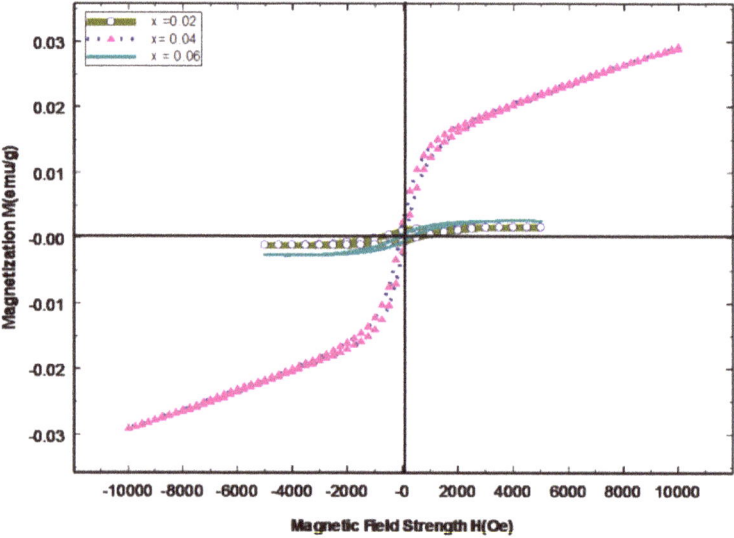

Figure 3.12 (a) M-H curves for $Zn_{1-x}V_xO$ (x=0.02, 0.04, 0.06) samples

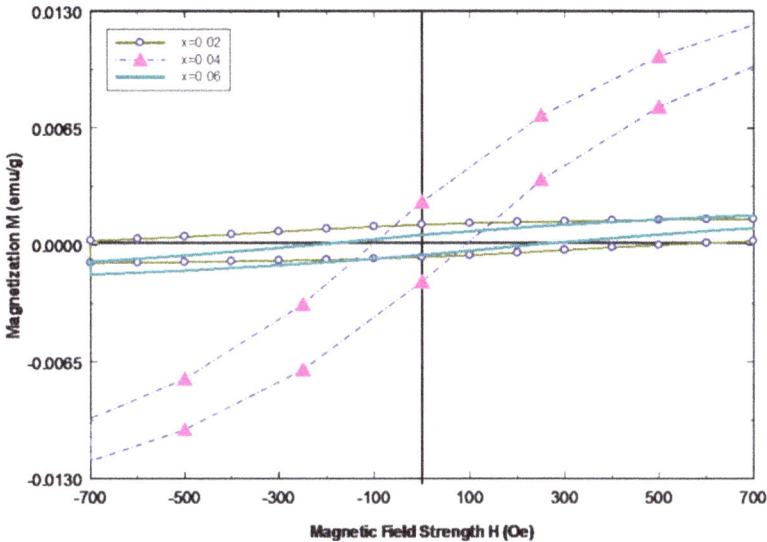

Figure 3.12 (b) Enhanced view of the hysteresis loop of $Zn_{1-x}V_xO$ (x=0.02, 0.04, 0.06) samples

Table 3.14 Magnetic parameters for the grown sample $Zn_{1-x}V_xO$ (x=0.02, 0.04, 0.06)

Concentrations	Coercivity (G)	Saturation Magnetization Ms(emu/g) $(x10^{-3})$	Retentivity (emu/g) $(x10^{-3})$	Remanence Ratio $(x10^{-3})$
x=0.02	900	0.0016	0.0011	0.68
x=0.04	090	0.0300	0.0029	0.09
x=0.06	210	0.0026	0.0006	0.23

3.4.4   Magnetic studies of $Zn_{1-x}Ni_{x/2}V_{x/2}O$ (x=0.02, 0.04, 0.06)

Ferromagnetic property of the grown samples were studied using vibrating sample magnetometery (VSM) of model Lakeshore VSM 7410 that works on Faraday's law of induction. Initially, the constant magnetic field is set and the sample is allowed to vibrate. The signal received from the probe is translated into a value for the magnetic moment of the sample. The magnetic field is then varied over a given range and a plot of magnetization (M) versus magnetic field strength (H) is generated. The resultant graph is shown as figures 3.13 (a) and (b) for the three concentrations. Room temperature ferromagnetic behavior.

was observed for all three concentrations x=0.02, 0.04, 0.06 through hysteresis curves. It confirms the doping of transition metal ions Ni and V remains successful. The M-H plots have authenticated the signature of the charge ordering and its corresponding magnetic states. The saturation magnetization of the three concentrations has an increasing trend with respect to increase in concentration it ranges from 6.0183 M_s to 15.3820 M_s as shown in Table 3.15. It is observed that as the transition metal ions Ni and V is low in concentration in x=0.02 samples we could possibly observe a weak ferromagnetic behavior. The ferromagnetic property increases with the increase in the concentration of Ni and V. These observations indicate that Ni^{2+} ions and V^{2+} ions systematically substituted for Zn sites without changing the wurtzite structure.

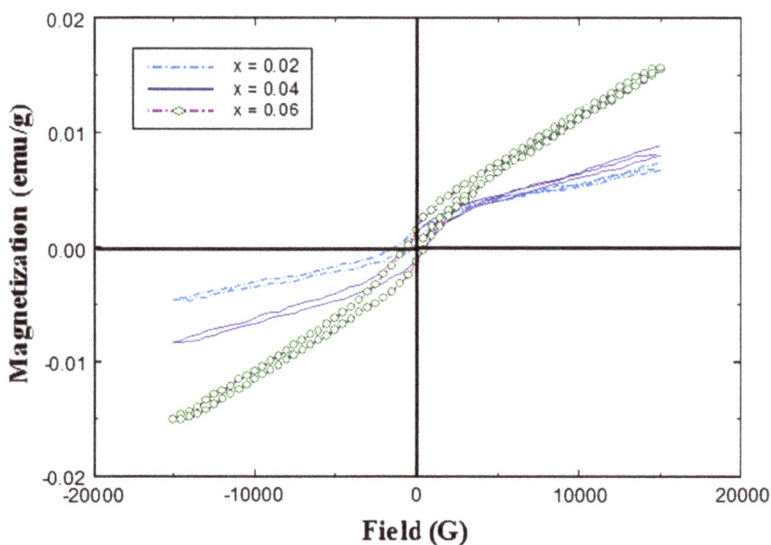

Figure 3.13 (a) M-H curves for $Zn_{1-x}Ni_{x/2}V_{x/2}O$ (x=0.02, 0.04, 0.06) samples

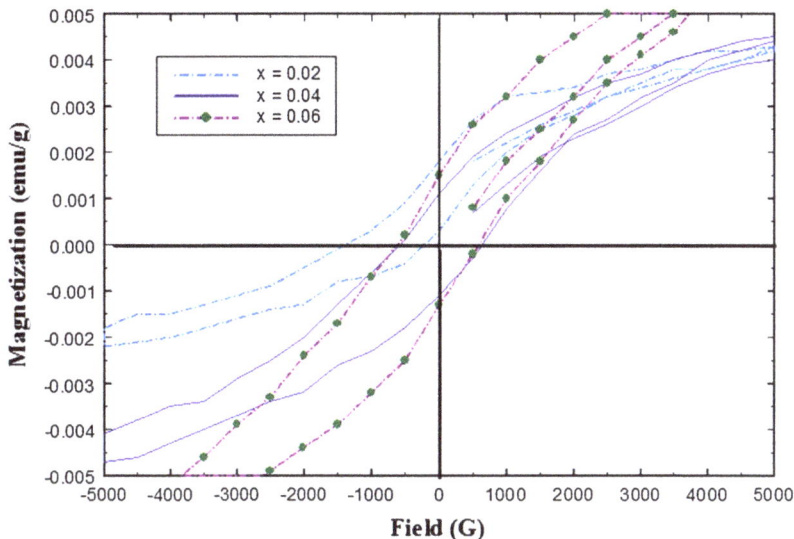

*Figure 3.13 (b) Enhanced view of the hysteresis loop of $Zn_{1-x}Ni_{x/2}V_{x/2}O$
(x=0.02, 0.04, 0.06) samples*

Table 3.15 M-H hystreis measurements from VSM

Concentrations	Coercivity (G)	Saturation Magnetization Ms(emu/g) $(x10^{-3})$	Retentivity (emu/g) $(x10^{-3})$
x=0.02	567	6.0183	0.7668
x=0.04	607	8.6192	1.0861
x=0.06	594	15.3820	1.3927

An outline of all the calculated values were shown in table 3.15 for the magnetic measurements of the grown samples. Table 3.15 may give an idea regarding the variation ofdopants and along with its variation inconcentrations.

Table 3.16 Corelative table of grown samples from VSM

Sample	Composition	Coercivity(G)	Saturation Magnetization, M_s(emu/g) x10^{-3}	Retentivity (emu/g) x10^{-3}
$Zn_{1-x}Ti_xO$	X= 0.01	278.56	5.4856	0.8563
	X= 0.02	280.43	6.9929	1.7261
	X= 0.03	356.08	8.0644	2.7222
$Zn_{1-x}Fe_xO$	X= 0.02	317.97	15.083	1.816
	X= 0.04	481.46	18.537	2.666
	X= 0.06	273.04	23.031	3.631
$Zn_{1-x}V_xO$	X= 0.02	900	0.0016	0.0011
	X= 0.04	090	0.0300	0.0029
	X= 0.06	210	0.0026	0.0006
$Zn_{1-x}Ni_{x/2}V_{x/2}O$	X= 0.02	567	6.0183	0.7668
	X= 0.04	607	8.6192	1.0861
	X= 0.06	594	15.3820	1.3927

References

[1] Krithiga R., Chandrasekaran G., J. Mat. Sci. doi:10.1016/j.jcrysgro.2009.08.033 (2011) https://doi.org/10.1016/j.jcrysgro.2009.08.033

[2] Kuznetsov A.S., Lu Y.-G., Turner S., Shestakov M.V., Tikhomirov V.K., Kirilenko D., Verbeeck J., Baranov A.N., Moshchalkov V.V., Opt. Mater. Express 2(6), 723-734 (2012). https://doi.org/10.1364/OME.2.000723

[3] Luo J.T., Zhu X.Y., Fan B., Zeng F., Pan F., J. Phys. D: Appl. Phys. 42, 115109 (2009). https://doi.org/10.1088/0022-3727/42/11/115109

[4] Samavati A., Nur H., Ismail A.F. and Othaman Z., J. Alloys Compd., 671, 170-176 (2016). https://doi.org/10.1016/j.jallcom.2016.02.099

[5] Saravanan R., Grain software, www.saraxraygroup.net.

[6] Singh S., Rama N., Sethupathi K., Ramachandra Rao M.S., J. Appl. Phys. 103(7), 07D108-1-07D108-3, (2008). https://doi.org/10.1063/1.2834443

[7] Wood D.L., Tauc J., Phys. Rev. B5, 3144 (1972). https://doi.org/10.1103/PhysRevB.5.3144

Transition Metal Doped Spintronics Materials
Materials Research Foundations **139** (2023)

Materials Research Forum LLC
https://doi.org/10.21741/9781644902257

Chapter 4

Charge Density Analysis

Abstract

The origin of charge ordering can be understood from the accurate electronic structure inside the unit cell and this can be achieved by constructing the charge density from structure factors using the best possible mathematical model like maximum entropy method (MEM). This method was successfully introduced to X-ray crystallography for constructing charge density in the unit cell by Collins (Collins, 1982) in 1982. MEM infers electron densities in such a way that they provide the maximum variance of structure factors, F_c (MEM), within errors in observed structure factors (F_o). MEM gives very accurate and less noisy density maps than the Fourier synthesis. In the present work, the computation of the charge density is done using the software PRIMA which employs MEM technique and the resultant density is plotted with the help of visualization software VESTA. Studies using MEM in understanding different types of bonding and the interior electronic details of many materials are available in literature. Another advantage of this method is the clear visualization of bonding nature and the distribution of electrons in the valance region with more accuracy.

Keywords

MEM, Charge Density, Charge Accumulation, (100) Plane, (110) Plane, Ionic Nature, Bond 1 and 2

4.1 Introduction

The origin of charge ordering can be understood from the accurate electronic structure inside the unit cell and this can be achieved by constructing the charge density from structure factors using the best possible mathematical model like maximum entropy method (MEM). This method was successfully introduced to X-ray crystallography for constructing charge density in the unit cell by Collins (Collins, 1982) in 1982. MEM infers electron densities in such a way that they provide the maximum variance of structure factors, F_c (MEM), within errors in observed structure factors (F_o). MEM gives very accurate and less noisy density maps than the Fourier synthesis (McCusker et al., 1999). In the present work, the computation of the charge density is done using the software PRIMA (Ruben et al., 2004, Izumi, Dilanian 2002,) which employs MEM technique and the resultant density is plotted with the help of visualization software VESTA (Momma, Izumi

2006). Studies using MEM in understanding different types of bonding and the interior electronic details of many materials are available in literature (e.g., (Yamamura et al., 1968; Gilmore, 1996; Livesey and Skilling, 1985; Sakata and Sato, 1990; De Vries et al. 1994; Kajitani et al., 2001; Saravanan et al., 2004; Israel et al., 2003; Syed Ali et al., 2006; Akilan et al., 2014)). Another advantage of this method is the clear visualization of bonding nature and the distribution of electrons in the valance region with more accuracy.

4.2 $Zn_{1-x}Ti_xO$ (x=0.01, 0.02, 0.03)

The methodology of MEM (maximum entropy method) electron densities were discussed in chapter I. The three dimensional electron densities were shown in figures 4.1 (a), (b) and (c). To understand the atomic occupancy better in the unit cell, the unit cell was partitioned on (110) plane which shows the bonds between zinc and oxygen atoms. Surprisingly the bonding between the oxygen positioned at $(-x, -y, z+0.5)$ and zinc atom exhibits covalent nature (bond 1) and the bonding between oxygen (at $-y, x-y, z$) and zinc exhibits ionic nature (bond 2). The contour lines are shown in figures 4.2 (a), (b) and (c), clearly shows the covalent electron densities in bond 1 and ionic bonding in bond 2. The orientation of the contour lines indicates that the bonding actually tends to be more covalent than ionic. This may be due to the large electronegativity of the oxygen atom. The iso-surface of zinc atoms increases with increase in the concentration of Titanium dopants as evident from figures 4.2 (a), (b) and (c) respectively. The 1D electron density profiles depicting the bonding between Zn and O atom along the directions (at $-x, -y, z+0.5$) and (at $-y, x-y, z$) are shown in figures 4.3 (a), (b) and (c) respectively. The numerical values of mid-bond electron densities were given in Table 4.1. The magnitudes of the charge density at the mid bond position ranges from 0.54 to 0.74 $e/\text{Å}^3$. It exhibits the bonding to be covalent with the increasing electron density at the mid bond position along with the increasing concentration of the Ti. A similar trend is observed between Zn and O (at $-y, x-y, z$), as shown in figure 4.4 (a), (b) and (c) gives the magnitudes of the charge density at the mid position ranging from 0.48 to 0.96 $e/\text{Å}^3$. These values show typically the bonding to be ionic.

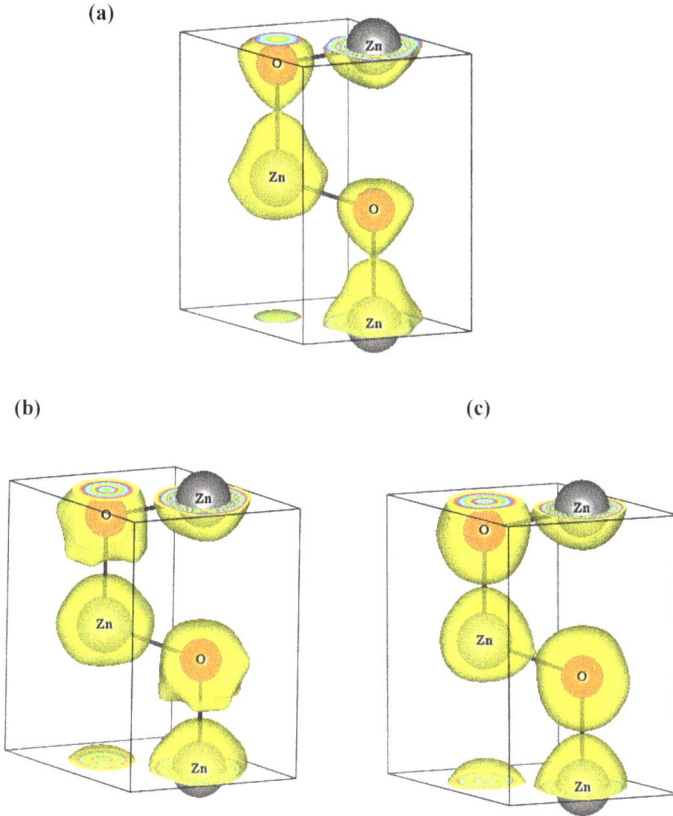

Figure 4.1. Three-dimensional electron density distributions of $Zn_{1-x}Ti_xO$ for (a) x = 0.01, (b)x=0.02 and (c) x=0.03) with iso surface level 0.65 e/$\mathring{A}^3$

(a)

(b)

(c)

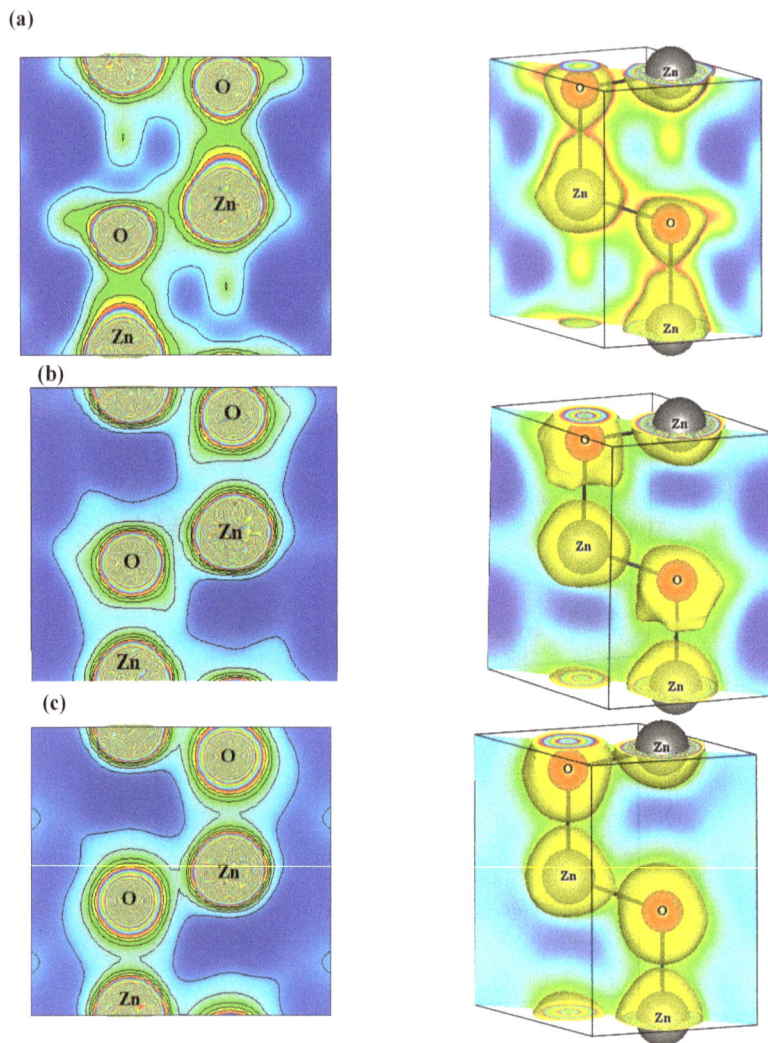

Figure 4.2. Two-dimensional electron density distributions of (110) plane in contour level 0-2 e/Å³ with interval of 0.3 e/ Å³ for ((a) $Zn_{0.99}Ti_{0.01}O$ (b) $Zn_{0.98}Ti_{0.02}O$ and (c) $Zn_{0.97}Ti_{0.03}O$

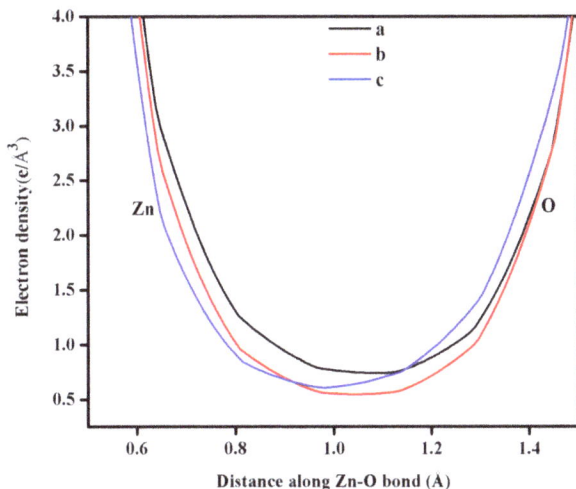

Figure 4.3. One-dimensional electron density distribution of $Zn_{1-x}Ti_xO$ ((a)x=0.01, (b)x=0.02 and (c) x=0.03) (bond1)

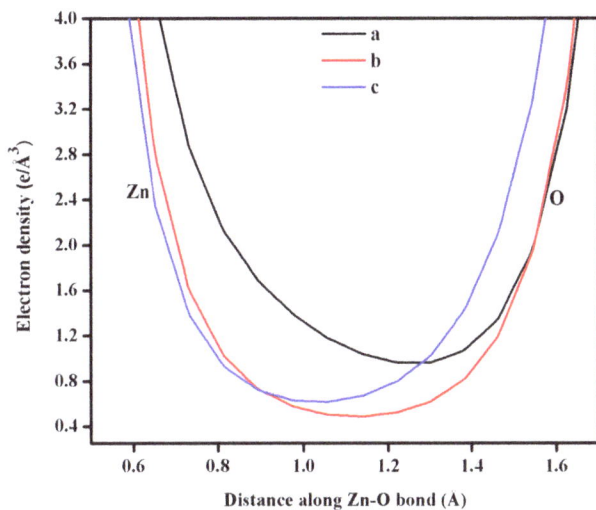

Figure 4.4. One-dimensional electron density distribution of $Zn_{1-x}Ti_xO$ ((a)x=0.01, (b)x=0.02 and (c) x=0.03) (bond2)

Table 4.1. Mid bond electron density values of $Zn_{1-x}Ti_xO$ for bond 1 and bond 2

$Zn_{1-x}Ti_xO$	Bond 1 (Horizontal axis)		Bond 2 (Vertical axis)	
	Bond length (Å)	Mid bond electron density (e/Å³)	Bond length (Å)	Mid bond electron density (e/Å³)
x = 0.01	1.9302	0.7423	2.1416	0.9624
x = 0.02	1.9397	0.5465	2.1065	0.4843
x = 0.03	1.9535	0.6089	2.0613	0.6128

4.3 $Zn_{1-x}Fe_xO$ (x=0.02, 0.04, 0.06)

In order to understand and throw more light on the inclusion of Fe atom in the Zn lattice site, the electron densities are computed on (110) plane, since the plane cuts both zinc and oxygen atoms at two different positions. The densities presented in the figures 4.5 (a)-(c), show the three-dimensional MEM electron density distributions of $Zn_{1-x}Fe_xO$ on the (110) plane respectively for x=0.02, 0.04 and 0.06, with the corresponding planes being shown in the unit cell. The contour ranges from 0.02 to 1.5 e/Å³ and the contour interval is 0.05 e/Å³, is maintained uniform for all the concentration levels to see the variation in different concentrations. These figures clearly reveal the enhancement of the charge distribution of the host Zn atom when the concentration of Fe is increased. In seen from these figures, that there is enhanced spatial distribution in the arrangement of charges at the Zn position compared to that of O. Residual electron densities was not observed, except in the regions allowed by symmetry and bonding. The two-dimensional iso-surfaces of $Zn_{1-x}Fe_xO$ are shown in figures 4.5 (d)-(f) for x=0.02, 0.04 and 0.06 respectively. From these figures we can classify the nature of the bonding between the Zn and oxygen atom into two types. The bond 1 indicate the interaction of contour lines between Zn and O (at –x, -y, z+0.5) showing the covalent nature of bonding. The bond 2 between Zn and O (at –y, x-y, z) is observed to be ionic, as the contour line between the atoms fades out.

It is also evident in the one-dimensional analysis, in figures 4.6 (a) and (b), where the high mid bond electron densities between Zn and O (bond 1) atoms lying between a range of 0.26 e/Å³ and 1.66 e/Å³ indicates the covalent nature of bonding. The low mid bond electron densities (bond 2) ranging from 0.31 e/Å³-0.70 e/Å³ are attributed for ionic nature of bonding. The mid bond electron densities for covalent and ionic bonding are comparable to the values obtained for $Zn_{1-x}V_xO$ (Akilan et al., 2014), $Zn_{1-x}Ni_xO$ (Syed Ali et al., 2011) and $Zn_{1-x}Co_xO$ (Ali et al., 2010). The mid bond electron densities between Zn atom and O atoms at bond 1 and bond 2 positions were tabulated in Table 4.2. The decrease in peak electron density with increase in the dopant concentration reveals the inclusion of Fe^{2+} ions in place of Zn^{2+} ions as seen from figure 4.6 (c). This trend gives more evidence about the inclusion of Fe^{2+} ions in Zn^{2+} ions along the ZnO host lattice since it have less electrons than Zn.

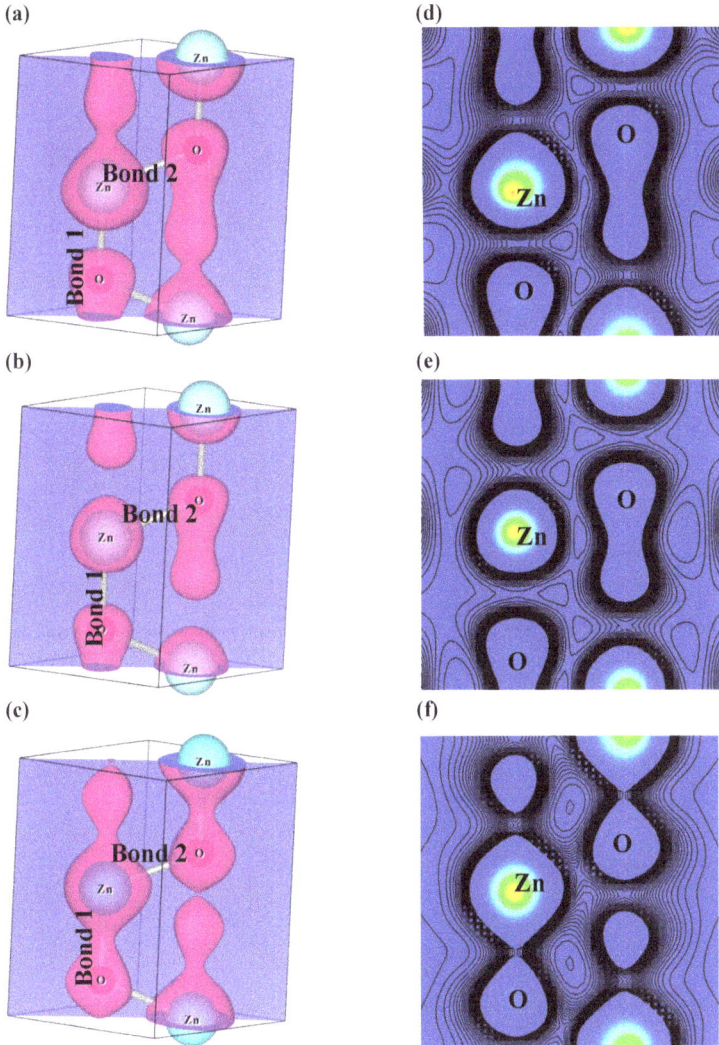

Figure 4.5 Three-dimensional charge density isosurfaces for $Zn_{1-x}Fe_xO$ with (110) plane shaded for a) x=0.02 b) x=0.04 c) x=0.06. Two-dimensional charge density distribution on (110) plane for $Zn_{1-x}Fe_xO$, d) x=0.02 e) x=0.04 f) x=0.06

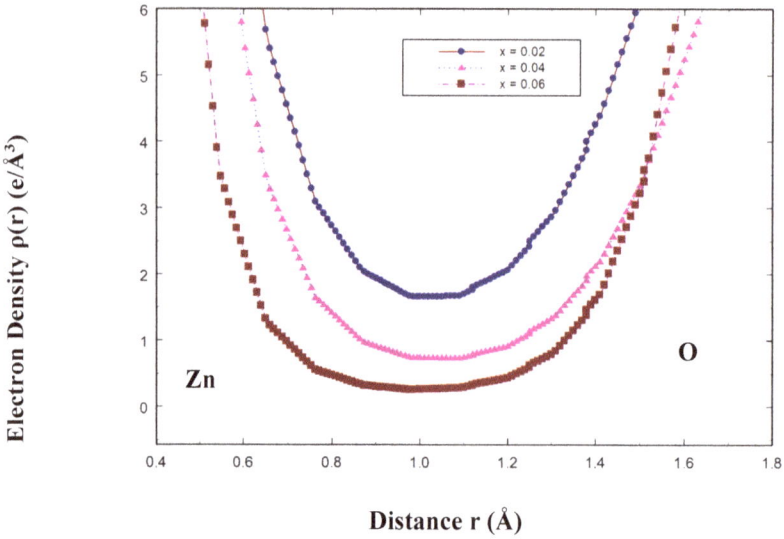

Figure 4.6 (a) One dimensional electron density profiles between Zn and O atoms at (-x, -y, z+1/2) for three compositions

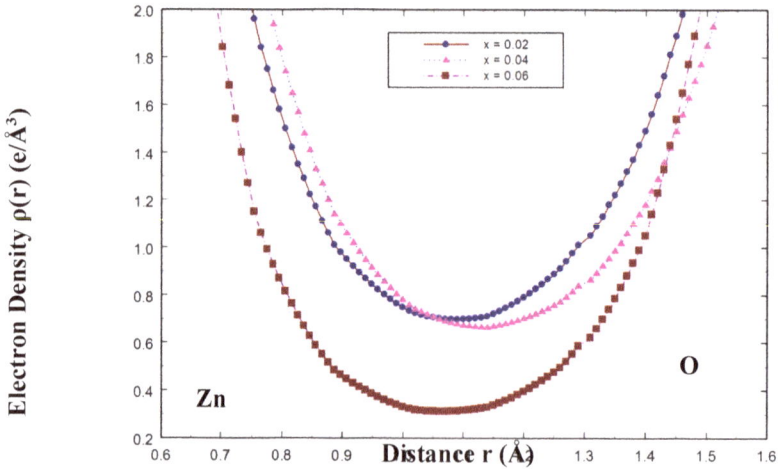

Figure 4.6 (b) One-dimensional electron density profiles between Zn and O atoms at (-y, x - y, z), for three compositions

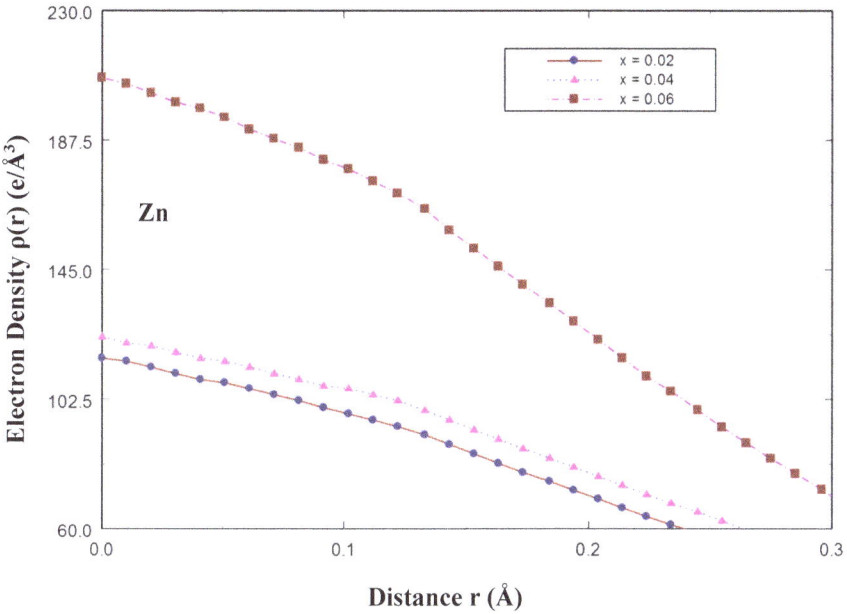

Figure 4.6 (c) Peak electron densities of $Zn_{1-x}Fe_xO$ for three compositions

Table 4.2 One-dimensional electron density values of $Zn_{1-x}Fe_xO$

Concentrations	Zn-O (bond1)		Zn-O (bond 2)	
	Distance r (Å)	**Density $\rho(r)$ (e/Å^3)**	**Distance r (Å)**	**Density $\rho(r)$ (e/Å^3)**
x= 0.02	1.01	1.66	1.09	0.70
x=0.04	1.07	0.73	1.12	0.67
x=0.06	0.99	0.26	1.06	0.31

4.4 $Zn_{1-x}V_xO$ (x=0.02, 0.04, 0.06)

The electronic charge distribution of $Zn_{1-x}V_xO$ with x=0.02, 0.04 and 0.06 is analyzed by maximum entropy method (MEM) (Denton et al., 1991, Kajitani et al., 2001) using the structure factors obtained from Rietveld (Saravanan, 2009; Rietveld, 1969) measurements. The MEM electron densities compiled from the experimental information are used for the visualization of the 3D electron density using the software VESTA (Momma and Izumi, 2008). The MEM electron density studies for various systems such as Na, V, Si, GaAs, Ge etc., have already been reported elsewhere (Saravanan, 2009; Momma and Izumi 2008; Saravanan et al., 2007; Saravanan et al., 2007; Saravanan et al., 2008; Saravanan et al., 2012).

The magnetic behaviour of a system depends on their charge arrangement and their involvement in the bonding process (Saravanan et al., 2011). Hence the electronic images obtained from the charge density studies were evaluated. In order to understand and throw more light on the inclusion of V atom in the Zn lattice site, the electron densities are computed on (110) plane, since the plane cuts both zinc and oxygen atoms at two different positions. The densities presented in the figures 4.7 (a)-(c), show the two-dimensional MEM electron density distributions of $Zn_{1-x}V_xO$ on the (110) plane respectively for x=0.02, 0.04 and 0.06, with the corresponding planes being shown in the unit cell. The contour ranges from 0.02 to 1.5 $e/Å^3$ and the contour interval is 0.05 $e/Å^3$, is maintained uniform for all the concentration levels to see the variation in different concentrations. These figures show the decrease in the concentration level as x increases from x=0.02 to 0.06. The three-dimensional iso-surfaces of $Zn_{1-x}V_xO$ are shown in figures 4.7 (d) to (f) for x=0.02, 0.04 and 0.06 respectively. From these figures we can classify the nature of the bonding between the Zn and oxygen atom into two types. The bond 1 indicate the interaction of contour lines between Zn and O (at –x, -y, z+0.5) showing the covalent nature of bonding. The bond 2 between Zn and O (at –y, x-y, z) is observed to be ionic, as the contour line between the atoms fades out.

It is also evident in the one-dimensional (1D) analysis, in figures 4.8 (a) and (b), where the high mid bond electron densities between Zn and O (bond 1) atoms lying between a range of 0.65 $e/Å^3$ and 0.72 $e/Å^3$ indicates the covalent nature of bonding. The low mid bond electron densities (bond 2) ranging from 0.15 $e/Å^3$ to 0.26 $e/Å^3$ are attributed for ionic nature of bonding. The mid bond electron densities for covalent and ionic bonding are comparable to the values obtained for $Zn_{1-x}Ni_xO$ (Saravanan et al., 2011) and $Zn_{1-x}Co_xO$ (Syed Ali et at., 2010). The mid bond electron densities between Zn atom and O atoms at bond 1 and bond 2 positions were tabulated in Table 4.3. The decrease in peak electron density with increase in the dopant concentration reveals the inclusion of V^{2+} ions in place of Zn^{2+} ions as seen from figure 4.8 (c). This trend gives more evidence about the inclusion of V^{2+} ions in Zn^{2+} ions along the ZnO host lattice since it have less electrons than Zn.

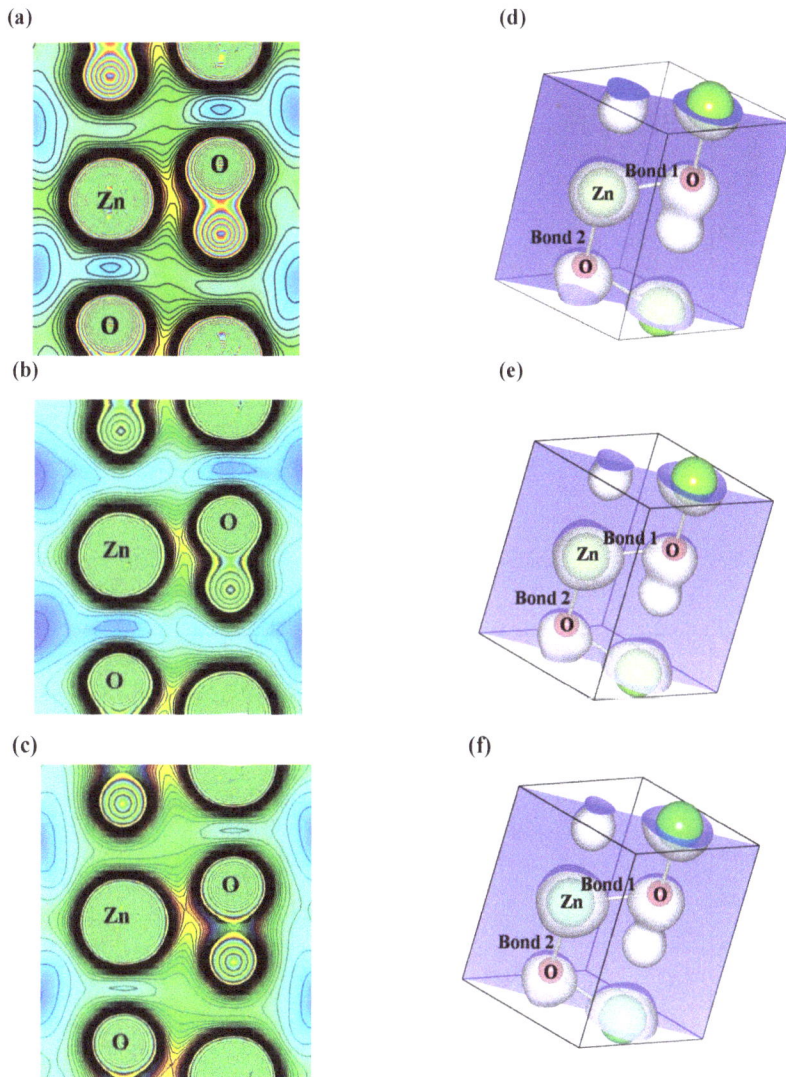

Figure 4.7 Two-dimensional charge density distribution on (110) plane for $Zn_{1-x}V_xO$, a) x=0.02 b) x=0.04 c) x=0.06. Three dimensional charge density isosurfaces for $Zn_{1-x}V_xO$ with (110) plane shaded for d) x=0.02 e) x=0.04 f) x=0.06

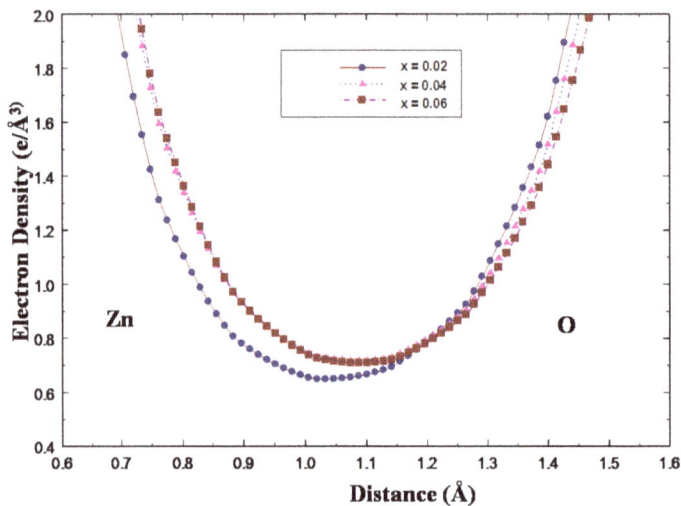

*Figure 4.8 (a) One-dimensional electron density profiles between Zn and O atoms at
(-x, -y, z+1/2) for three compositions*

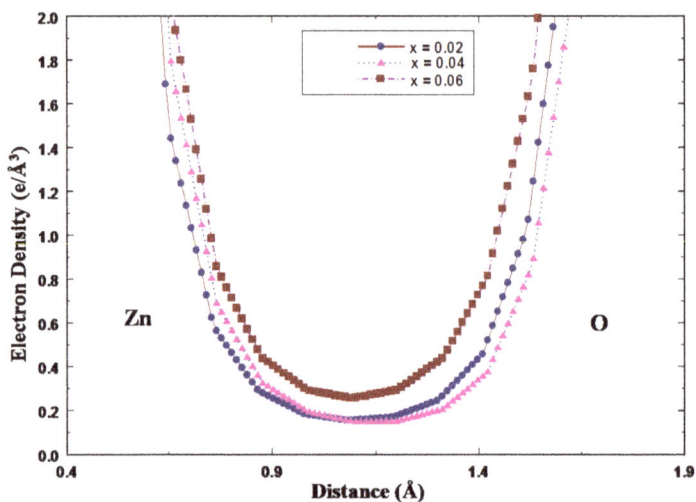

*Figure 4.8 (b) One-dimensional electron density profiles between Zn and O atoms at
(-y, x - y, z), for three compositions*

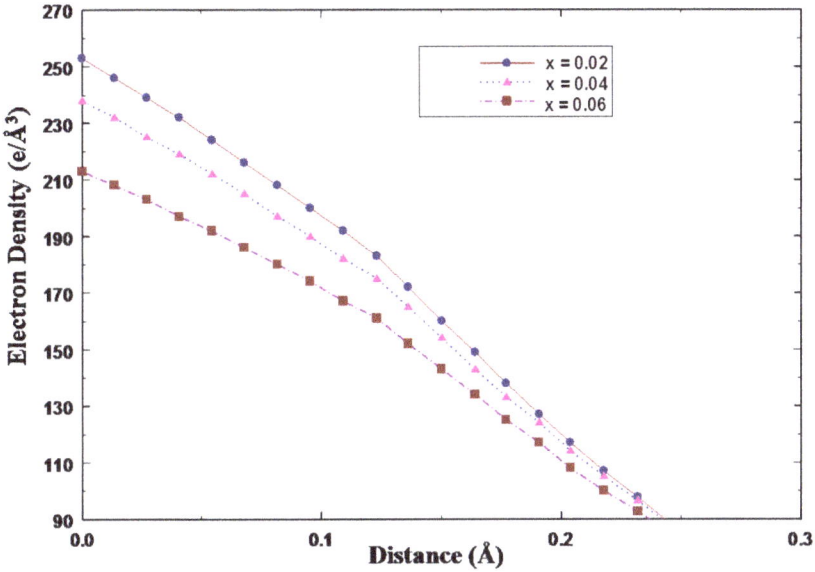

Figure 4.8 (c) Peak electron densities of $Zn_{1-x}V_xO$ for three compositions

Table 4.3 One-dimensional electron density values of $Zn_{1-x}V_xO$

Concentrations	Zn-O (bond1)		Zn-O (bond 2)	
	Distance r (Å)	Density $\rho(r)$ (e/Å³)	Distance r (Å)	Density $\rho(r)$ (e/Å³)
x= 0.02	1.03	0.65	1.19	0.15
x=0.04	1.08	0.72	1.14	0.15
x=0.06	1.09	0.71	1.08	0.26

4.5 $Zn_{1-x}Ni_{x/2}V_{x/2}O$ (x=0.02, 0.04, 0.06)

The charge density derived from Maximum Entropy Method (MEM) and it has been applied through PRIMA package and VESTA software. The charge density distribution in the unit cell was shown in figures 4.9 (a)-(c). The 3-D charge density is drawn at the iso-surface level of 1.26 e/Å3 and it can be understood that the influence of dopant concentration at the host lattice site. 2-D electron density maps were drawn on (110) plane which shows the bonds between zinc and oxygen atoms. Surprisingly the bonding between the oxygen positioned at (–x, -y, z+0.5) and zinc atom exhibits covalent nature (bond 1) and the bonding between oxygen (at –y, x-y, z) and zinc exhibits ionic nature (bond 2). The contour lines are shown in figures 4.9 (d)-(f), clearly shows the covalent electron densities in bond 1 and ionic bonding in bond 2. The 1D electron density profiles depicting the bonding between Zn and O atom along the directions (at $-x$, $-y$, $z+0.5$) and (at $-y$, x-y, z) are shown in figures 4.10 (a) and (b) respectively. The numerical values of mid-bond electron densities were given in Table 4.4. The low electron density regions in figure 4.10(a) saddles, with a little flattening at the mid bond positions. The magnitudes of the charge density at the mid bond position ranges from 0.24 to 0.93 e/Å^3. It exhibits the bonding to be covalent with the increasing electron density at the mid bond position along with the increasing concentration of the Ni and V. A similar trend is observed between Zn and O (at -y, x-y, z), as shown in figure 4.10 (b) gives the magnitudes of the charge density at the mid position ranging from 0.43 to 0.54 e/Å^3. These values show that typically the bonding to be ionic. Furthermore, confirmation regarding the addition Ni and V ions at the lattice site of Zn atoms is analyzed from the peak electron density presented in figure 4.10 (c). The figure explains the discussed ideas, when Ni and V dopant concentration increases it actually decreases the peak density at the lattice site of Zn due to the fact that Ni and V have less electrons or atomic radius than Zn atoms.

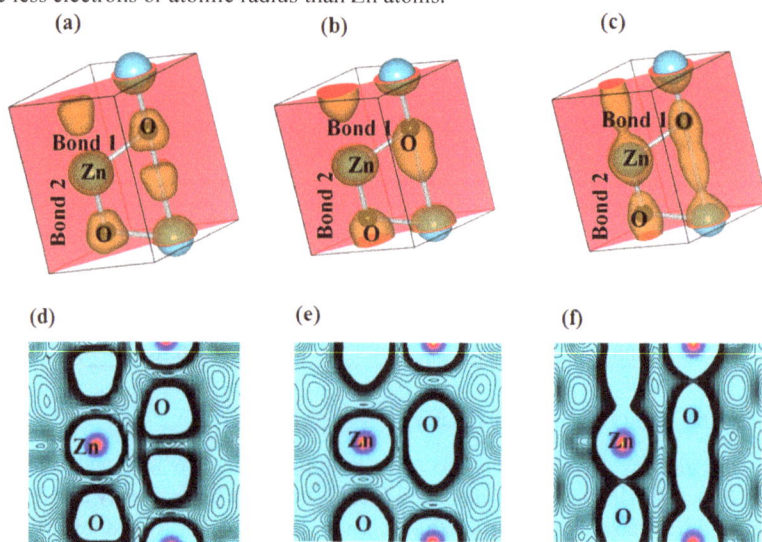

Figure 4.9 Three dimensional charge density isosurfaces for $Zn_{1-x}Ni_{x\,2}V_{x\,2}O$ with (110) plane shaded for a) x=0.02 b) x=0.04 c) x=0.06.
Two dimensional charge density distribution on (110) plane for $Zn_{1-x}Ni_{x\,2}V_{x\,2}O$, d) x=0.02 e) x=0.04 f) x=0.06.

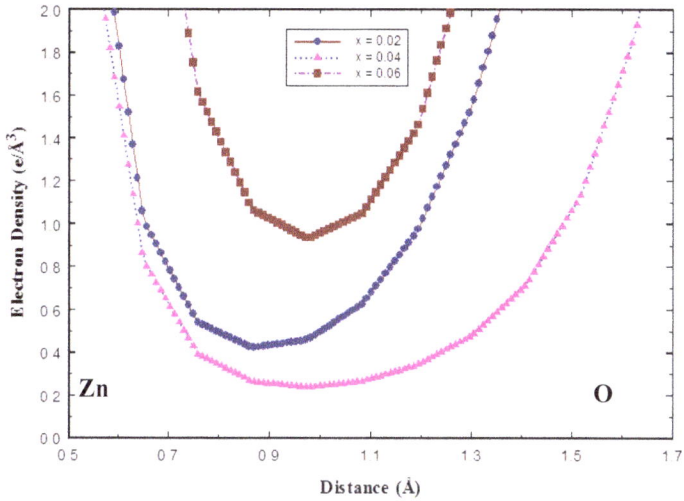

Figure 4.10 (a) One dimensional electron density profiles between Zn and O atoms at (-x, -y, z+1/2) for three compositions

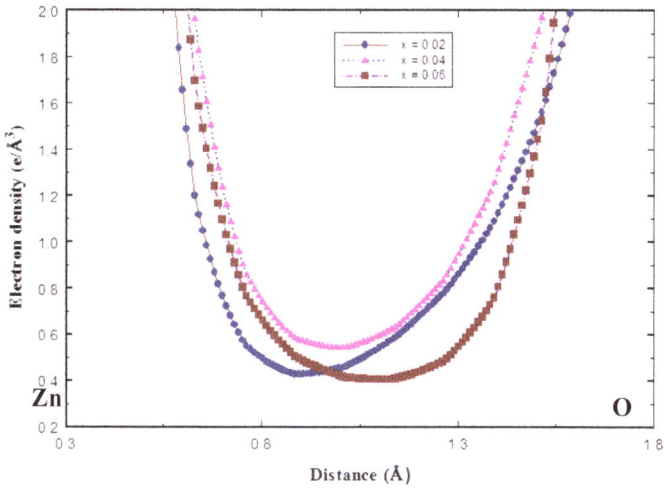

Figure 4.10 (b) One-dimensional electron density profiles between Zn and O atoms at (-y, x - y, z), for three compositions

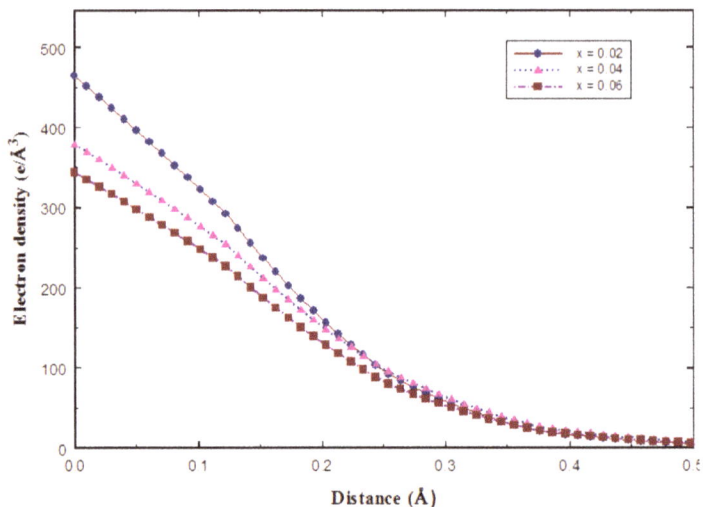

Figure 4.10 (c) Evaluation of peak electron densities for three compositions

Table 4.4. One-dimensional electron density values of $Zn_{1-x}Ni_{x/2}V_{x/2}O$ (x=0.02, 0.04, 0.06)

Concentrations	Zn-O (bond 1)		Zn-O (bond 2)	
	Distance r (Å)	Density $\rho(r)$ (e/Å³)	Distance r (Å)	Density $\rho(r)$ (e/Å³)
x=0.02	0.91	0.42	0.87	0.42
x=0.04	0.99	0.54	0.97	0.24
x=0.06	1.08	0.40	0.97	0.93

Comparing all the values of mid bond electron densities of the frown samples table 4.5 was depicted to understand the effect of various dopants.

Table 4.5 Comparative values for all dopants with their various concentrations

Sample	Composition	Bond 1(horizontal axis)			Bond 2 (vertical axis)		
		Bond Length	Mid bond electron density $(e/Å^3)$	Relative nature of bond	Bond Length	Mid bond electron density $(e/Å^3)$	Relative nature of bond
$Zn_{1-x}Ti_xO$	X= 0.01	1.93	0.74		2.14	0.96	
	X= 0.02	1.93	0.54		2.10	0.48	
	X= 0.03	1.95	0.60		2.06	0.61	
$Zn_{1-x}Fe_xO$	X= 0.02	1.01	1.66		1.09	0.70	
	X= 0.04	1.07	0.73		1.12	0.67	
	X= 0.06	0.99	0.26	Covalent Bond	1.06	0.31	Ionic bond
$Zn_{1-x}V_xO$	X= 0.02	1.03	0.65		1.19	0.15	
	X= 0.04	1.08	0.72		1.14	0.15	
	X= 0.06	1.09	0.71		1.08	0.26	
$Zn_{1-x}Ni_{x/2}V_{x/2}O$	X= 0.02	0.91	0.42		0.87	0.42	
	X= 0.04	0.99	0.54		0.97	0.24	
	X= 0.06	1.08	0.40		0.97	0.93	

References

[1] Akilan T., Srinivasan N., Saravanan R., J. Mater. Sci.: Mater. Electron. (2014) doi:10.1007/s10854-014-1957-4. https://doi.org/10.1007/s10854-014-1957-4

[2] Ali K.S.S., Saravanan R., Ac¸ıkgo¨z M., Cryst. Res. Technol. (2011) doi:10.1002/crat.201000387. https://doi.org/10.1002/crat.201000387

[3] Ali K.S.S., Saravanan R., Israel S., Acrkgoz M., Arda L., Phys. B (2010) doi:10.1016/j.physb.2010.01.036. https://doi.org/10.1016/j.physb.2010.01.036

[4] Ali K.S.S., Saravanan R., Israel S., Rajaram R.K., Bull. Mater. Sci. (2006) doi:10.1007/BF02704601. https://doi.org/10.1007/BF02704601

[5] Collins D.M., Nature. 298, 49 (1982). https://doi.org/10.1038/298049a0

[6] De Vries R.Y., Brils W.J., Feil D., Acta Crystallogr. A (1994) doi:10.1107/S0108767393012802. https://doi.org/10.1107/S0108767393012802

[7] Denton A.R., Ashcroft N.W., Vegard's law. Physical Review A, 43, 3161-3164 (1991). https://doi.org/10.1103/PhysRevA.43.3161

[8] Gilmore C.J., Acta Crystallogr. A (1996) doi:10.1107/ S0108767396001560.

[9] Israel S., Saravanan R., Srinivasan N., Rajaram R.K., J. Phys. Chem. Solids 64, 43 (2003). https://doi.org/10.1016/S0022-3697(02)00208-1

[10] Israel S., Saravanan R., Srinivasan N., Rajaram R.K., J. Phys.Chem. Solids (2003) doi:10.1016/S0022-3697(02)00208-1. https://doi.org/10.1016/S0022-3697(02)00208-1

[11] Izumi F., Dilanian R.A., Recent Research Developments in Physics (Transworld Research Network, Trivandrum, 2002), pp. 699-726.

[12] Kajitani T., Saravanan R., Ono Y., Ohno K., Isshiki M., Journal of Crystal Growth, 229,130-136 (2001). https://doi.org/10.1016/S0022-0248(01)01107-1

[13] Kajitani T., Saravanan R., Ono Y., Ohno K., Isshiki M., J. Cryst. Growth (2001) doi:10.1016/S0022-0248(01)01107-1. https://doi.org/10.1016/S0022-0248(01)01107-1

[14] Livesey A. K., Skilling J., Acta Crystallogr. A (1985) doi:10.1107/S0108767385000241. https://doi.org/10.1107/S0108767385000241

[15] McCusker L.B., Von Dreele R.B., Cox D.E., Lou¨er D., Scardi P., J. Appl. Cryst. (1999). doi:10.1107/S0021889898009856 https://doi.org/10.1107/S0021889898009856

[16] Momma K., Izumi F., IUCr Newslett. 7, 106 (2006).

[17] Momma K., Izumi F., VESTA: A three-dimensional visualization system for electronic and structural analysis. Journal of Applied Crystallography, 41, 653-658 (2008). https://doi.org/10.1107/S0021889808012016

[18] Rietveld H.M., Journal of Applied Crystallography, 2, 65-71 (1969). https://doi.org/10.1107/S0021889869006558

[19] Ruben A.D., Izumi F., Super-Fast Program PRIMA for the Maximum-Entropy Method (Advanced Materials Laboratory, National Institute for Materials Science. 1-1 Namiki, Tsukuba, Ibaraki Japan 305, 2004), p. 0044.

[20] Sakata M., Sato M., Acta Crystallogr. A (1990). doi:10.1107/S0108767389012377 https://doi.org/10.1107/S0108767389012377

[21] Saravanan R, Israel S., Phys. B (2004). doi:10.1016/j.physb.2004.07.014. https://doi.org/10.1016/j.physb.2004.07.014

[22] Saravanan R., Ann A.M.M., Jainulabdeen S., Physica B: Condensed Matter, 400, 16-21 (2007). https://doi.org/10.1016/j.physb.2007.06.010

[23] Saravanan R., Francis S., John Berchmans L., Chemical Papers, 66 (3), 226-234

(2012). https://doi.org/10.2478/s11696-011-0129-8

[24] Saravanan R., Israel S., Physica B 349, 220 (2004). https://doi.org/10.1016/j.physb.2004.07.014

[25] Saravanan R., Israel S., Rajaram R.K., Phys. B (2005). doi:10.1016/j.physb.2004.03.018. https://doi.org/10.1016/j.physb.2004.03.018

[26] Saravanan R., Physica Scripta 2009, 79, 048303. https://doi.org/10.1088/0031-8949/79/04/048303

[27] Saravanan R., Premarani M., Journal of Physics: Condensed Matter, 19, 266221 (2007). https://doi.org/10.1088/0953-8984/19/26/266221

[28] Saravanan R., Syed Ali K.S., Israel S., Pramana, 4, 679-696 (2008). https://doi.org/10.1007/s12043-008-0029-9

[29] Saravanan, Israel S., Rajaram R.K., Physica B 363, 166 (2005). https://doi.org/10.1016/j.physb.2005.03.018

[30] Saravanan, R.; Syed Ali, K.S.; Acrkgoz, M., Crystal Research and Technology, 46, 41-47 (2011) https://doi.org/10.1002/crat.201000387

[31] Syed Ali, K.S., Saravanan R., Israel S., Acrkgoz M., Arda L., Physica B, 405, 1763-1769 (2010). https://doi.org/10.1016/j.physb.2010.01.036

[32] Yamamura Y., Takata M., Sakata M., Sugawara Y., J. Phys. Soc. Jpn. (1968). doi:10.1143/JPSJ.67.4124 https://doi.org/10.1143/JPSJ.67.4124

Transition Metal Doped Spintronics Materials   Materials Research Forum LLC
Materials Research Foundations **139** (2023)   https://doi.org/10.21741/9781644902257

Chapter 5

Conclusions

Abstract

The investigative results and discussion obtained from previous chapters has enlightened some of the oxide based dilute magnetic materials. The grown sample establishes the relationship between local structures, morphological behaviour, semiconducting properties and magnetic interactions. A few important points were discussed in this chapter.

Keywords

ZnO, Vi, Ti, Ni, Fe

5.1 $Zn_{1-x}Ti_xO$ (x= 0.01, 0.02 and 0.03)

$Zn_{1-x}Ti_xO$ powders with varying concentrations as x= 0.01, x= 0.02 and 0.03 was prepared by solid state reaction method. The XRD analysis show left shift of Bragg peaks indicating Ti^{4+} ion being incorporated into the lattice site of the host ZnO. The crystallite size and average particle size increases with the dopant concentration. The samples show an enhancement of ferromagnetic behaviour, with addition of Ti in the host lattice. The optical band gap also shows an increasing trend with the concentration. Thus, the addition of Ti induces changes in ferromagnetic and semiconducting nature of the sample. Similarly, there was an observable red shift in the emission spectra with respect to the concentration.

5.2 $Zn_{1-x}Fe_xO$, (x= 0.02, 0.04 and 0.06)

The system $Zn_{1-x}Fe_xO$ is grown with various doping concentration x=0.02, x=0.04 and x=0.06 by standard solid state reaction method. The structure of Fe doped zinc oxide has been studied in terms of structure and electron density distributions and the effect of substitutional impurity (Fe) in the host lattice (ZnO) in the system were analyzed. The MEM analyses have been employed to mine maximum information from X-ray data. In the present study, the mid bond electron density between Zinc and Oxygen at two different positions have been evaluated to determine the nature of bonding. Covalent nature of bonding was observed at bond 1 position and ionic nature of bonding was observed at bond 2 position. The room temperature ferromagnetism was observed for all the three compositions. The micro structure measurements were carried out using field emission scanning electron microscopy (FESEM) and EDAX.

5.3 $Zn_{1-x}V_xO$, (x= 0.02, 0.04 and 0.06)

$Zn_{1-x}V_xO$ is grown with various doping concentration x=0.02, x=0.04 and x=0.06 by standard solid state reaction method. The structure of vanadium doped zinc oxide has been studied in terms of structure and electron density distributions and the effect of substitutional impurity (V) in the host lattice (ZnO) in the system was analyzed. The MEM analyses have been employed to mine maximum information from X-ray data. In the present study, the mid bond electron density between Zinc and Oxygen at two different positions have been evaluated to determine the nature of bonding. Covalent nature of bonding was observed at bond 1 position and ionic nature of bonding was observed at bond 2 position. The room temperature ferromagnetism was observed for all the three compositions. Among the three compositions the ferromagnetic properties of the three compositions concentration x= 0.04 remains to have high saturation point. The micro structure measurements were carried out using field emission scanning electron microscopy (FESEM) and EDAX.

5.4 $Zn_{1-x}Ni_{x/2}V_{x/2}O$, (x= 0.02, 0.04 and 0.06)

The transition metal ions Ni and V were doped in semiconducting material ZnO through solid state reaction method. The structural parameters and electron densities of the grown system with respect to their compositions were calculated. The study reveals the inclusion of Ni and V dopants on the Zn lattice. Electron density studies reveal the material to be polar covalent associated with covalent and ionic behavior. Quantitative measurement on ferromagnetic property was carried out through vibrating sample magnetometry (VSM). Room temperature ferromagnetism was confirmed for all three compositions. ICP-AES studies confirm the transition metal dopants of Ni and V in ZnO lattice. Direct band gap of the material was measured through UV-Vis spectra. The grown system may be suitable candidate for DMS material which have great uses in spintronic field and other related technological fields.

About the Author

Dr. Ramachandran Saravanan has been associated with the Department of Physics, The Madura College, affiliated with the Madurai Kamaraj University, Madurai, Tamil Nadu, India from the year 2000. He is the head of the Research Centre and PG department of Physics. He worked as a research associate during 1998 at the Institute of Materials Research, Tohoku University, Sendai, Japan and then as a visiting researcher at Centre for Interdisciplinary Research, Tohoku University, Sendai, Japan up to 2000.

Earlier, he was awarded the Senior Research Fellowship by CSIR, New Delhi, India, during Mar. 1991 - Feb.1993; awarded Research Associateship by CSIR, New Delhi, during 1994 – 1997. Then, he was awarded a Research Associateship again by CSIR, New Delhi, during 1997- 1998. Later he was awarded the Matsumae International Foundation Fellowship in1998 (Japan) for doing research at a Japanese Research Institute (not availed by him due to the simultaneous occurrence of other Japanese employment).

He has guided twelve Ph.D. scholars as of 2018, and about six researchers are working under his guidance on various research topics in materials science, crystallography and condensed matter physics. He has published around 150 research articles in reputed Journals, mostly International, apart from around 50 presentations in conferences, seminars and symposia. He has also guided around 60 M.Phil. scholars and an equal number of PG students for their projects. He has attracted government funding in India, in the form of Research Projects. He has completed two CSIR (Council of Scientific and Industrial Research, Govt. of India), one UGC (University Grants Commission, India) and one DRDO (Defense Research and Development Organization, India) research projects successfully and is proposing various projects to Government funding agencies like CSIR, UGC and DST.

He has written 12 books in the form of research monographs including; "Experimental Charge Density - Semiconductors, oxides and fluorides" (ISBN-13: 978-3-8383-8816-8; ISBN-10:3-8383-8816-X), "Experimental Charge Density - Dilute Magnetic Semiconducting (DMS) materials" (ISBN-13: 978-3-8383-9666-8; ISBN-10: 3-8383-9666-9) and "Metal and Alloy Bonding - An Experimental Analysis" (ISBN -13: 978-1-4471-2203-6). He has committed to write several books in the near future.

His expertise includes various experimental activities in crystal growth, materials science, crystallographic, condensed matter physics techniques and tools as in slow evaporation, gel, high temperature melt growth, Bridgman methods, CZ Growth, high vacuum sealing etc. He and his group are familiar with various equipment such as: different types of cameras; Laue, oscillation, powder, precession cameras; Manual 4-circle X-ray diffractometer, Rigaku 4-circle automatic single crystal diffractometer, AFC-5R and AFC-7R automatic single crystal diffractometers, CAD-4 automatic single crystal diffractometer, crystal pulling instruments, and other crystallographic, material science related instruments. He and his group have sound computational capabilities on different types of computers such as: IBM – PC, Cyber180/830A – Mainframe, SX-4 Supercomputing system – Mainframe. He is familiar with various kind of software related to crystallography and materials science. He has written many computer software programs himself as well. Around twenty of his programs (both DOS and GUI versions) have been included in the SINCRIS software database of the International Union of Crystallography.